Physics and Music

Kinko Tsuji · Stefan C. Müller

Physics and Music

Essential Connections and Illuminating Excursions

 Springer

Kinko Tsuji
Shimadzu Europa GmbH (retired)
Duisburg, Germany

Stefan C. Müller
Professor emeritus, Institute of Physics
Otto von Guericke University Magdeburg
Magdeburg, Germany

ISBN 978-3-030-68678-9   ISBN 978-3-030-68676-5   (eBook)
https://doi.org/10.1007/978-3-030-68676-5

This Springer imprint is published by the registered company Springer Nature Switzerland AG
The registered company address is: Gewerbestrasse 11, 6330 Cham, Switzerland

Preface

Both authors, KT and SCM, are scientists who love music and also play music together. Probably for that reason, SCM gave a lecture "Physik der Musik (Physics of Music)" at the University of Magdeburg in 2014 and 2015. Later, KT wrote down his lecture notes (in German) and considered a future publication. The main contents were "Unser modernes Tonsystem" (Our modern tonal system), "Physik der Musikinstrumente" (Physics of musical instruments), and "Sinfonieorchester und Raumakustik" (Symphony orchestra and room acoustics).

Soon after, we planned to write an English book on the basis of the lecture notes and discussed the options with the publisher. The publisher suggested to add some more topics and to write a book with the title *Physics and Music* instead of "Physics of Music". It should include, in addition, history, physiology, psychology, philosophy, and anecdotes. Initially we hesitated to accept the suggestion, because it covers a very wide field for which we are not specialists.

After a while, however, we decided to write a book according to these suggestions with our hope and ambition to build a bridge between scientists and musicians. In fact, we know a lot of scientists who love music and who play instruments. We also have contact with many professional musicians. It would be nice if these scientists and musicians communicate more often and possibly play and hear music together.

Small Stefan alias Wandering Musician

Kinko made a sketch from
a photograph by Erich Auerbach
in „Images of Music"
Könemann, 1996

The physics described in this book is in substantial parts based on the works of two persons: the German physicist and physician Hermann von Helmholtz and the British scientist Lord Rayleigh. The books *"Die Lehre von den Tonempfindungen als physiologische Grundlage für die Theorie der Musik"* *(On the Sensations of Tone as a Physiological Basis for the Theory of Music)* of Helmholtz, and *"The Theory of Sound"* of Rayleigh were published in 1863 and 1877, respectively.

On the side of music, pioneering and epoch-making contributions are quite multi-facetted. However, we like to name two personalities: Jean-Philippe Rameau and Johann Sebastian Bach. Rameau wrote a book *"Traité de l'harmonie réduite à ses principes naturels" (Treatise on Harmony reduced to its natural principles)* in 1722, and Bach created "Die Kunst der Fuge" (The Art of Fugue), reaching the most eminent peak of the counterpoint in 1751.

Moreover, during writing this book, we have learned a lot from the following four publications:

- E. Meyer and E.-G. Neumann, *Physikalische und Technische Akustik* (Vieweg, Braunschweig, 1974)
- J. Meyer, *Acoustics and the Performance of Music* (Springer Science+Business Media, Berlin, 2009)
- K. Winkler (Einführung), *Die Physik der Musikinstrumente* (Spektrum der Wissenschaft, Heidelberg, 1988)

- N. H. Fletcher and T. D. Rossing, *The Physics of Musical Instruments* (Springer Science+Business Media, Berlin, 1998)

We will start in our first chapter with a treatise of historical developments which have led to our modern concepts and understanding of music and its physical implications. History reaches far back in revealing evidence of musical activities. Flutes with finger holes carved of bone were used already in neolithic times, an early intuitive understanding behind the created sound. Old tribes have developed their specific singing tunes. Music of the ancient Mesopotamians (mostly for vocalists) involved both religious and social aspects.

Music as an ancient means of people to communicate with each other has been always linked with attempts to understand its basic structure and function. Musical sounds and instruments have been, in fact, an integral feature of the earliest activities of mankind. And efforts to understand music as such or ways to produce music belong to very early endeavors of knowledgeable members of human communities.

In the following three chapters, we introduce the European modern tonal system. Music is an art made from the realm of sounds, mainly characterized by melody, harmony, and rhythm. "Sound" is a very complex quantity, it is generated by the distribution of acoustic frequencies that can produce singular tones, sound with many overtones depending on the nature of the instrument, colored and accentuated noise as created by drums, or bangs from a timpani. In its structure, a melody often is following a tonal sequence obeying specific laws (Chap. 2). Here the basic "grammar" of the musical notes is also explained.

Tonal sequences will be described and discussed on physical and mathematical grounds in Chap. 3. In ancient Greece, Pythagoras, Philolaus, Euklid, and some others already provided a mathematical construction of a circle of fifths (quints) based on the frequency ratio 3:2. During the following centuries, apart from many versions of pentatonic tone scales, medieval religious preferences in church music modeled diatonic scales (using a mixture of full and half tones), until from the year 1300 on the "just intonation" became popular. It is based on the harmonic series generated by instruments, in particular "linear" ones like the chord or the flute. The just intonation proved to be more suited when certain different instruments played together.

However, there are basic differences in frequencies of harmonics depending on the instruments. As described in Chap. 4, several steps were undertaken further on to create an optimum of frequency relationships, an important one being the well-tempered scale used in the Baroque period. Today the "equal temperament" is mostly used, a mathematically created scale with logarithmically equal intervals between half tones. It is a compromise, making it possible that a piano and a violin (for instance) can find an agreeable way to play together. The compromise uses, among other aspects, the fact that the frequency resolution of our ear is limited.

In Chap. 5, we travel through countries and continents surrounding Europe: Arabic countries, India, East Asia, Oceania, Africa, and the Americas. We explore and analyze musical sound systems and their empiric physical laws, as they have developed over time in different cultures.

A large portion of the book will deal with musical instruments: their sound characteristics, their structure, physical mechanisms of sound creation and emission, tonal capacities and limitations, modalities of appropriate use, and more. A guideline is set according to the families of instruments commonly used in symphony orchestras and, for all examples, cross-links to "relatives" from different cultural environments will be considered. They are found in Chaps. 6–8. As selected instruments, we introduce the recorder, the transverse flute, the oboe, and some more woodwind and brass instruments. Furthermore, we discuss strings, in particular the violin, as well as several percussion instruments.

What happens, when musicians get together? They create music (Chap. 9), they play music together, and they form an orchestra (Chap. 10).

Music without audience is difficult to imagine. In Chap. 11, we try to explain how our hearing mechanism works, though honestly speaking, there are many things we do not yet understand and which remain hidden in a huge black box. As next themes, we choose absolute and relative pitches, as well as auditory illusions.

The acoustic properties of the room or hall have a strong influence on sound perception. In fact, room acoustics has become a separate field of acoustics, where interdisciplinary exchange between scientists, architects, and artists (and financial agencies) takes place. We will be carried into the domains of nonlinear mathematics and physics to grasp the complexity of a sound field in a concert hall (Chap. 12).

Immediately after we started this project, we noticed that many complications in music theory stem from the line system for writing musical notes: The intervals of lines are not equal, although the intervals written on the paper of two adjacent lines are equal. It is completely irrational to draw lines of the same distance for two whole tones or 1 whole tone + 1 half tone. A few years ago KT developed a method to interpret musical pieces by the dynamical means of the phase plot. We will show in Chap. 13, how musical pieces and various musical scales can be visualized as phase portraits.

Finally, we debate the relationship between music, our mind, and our society (Chap. 14). We try to understand the differences between a major key and a minor key in musical passage. Which factors cause sadness or happiness, for example? We also analyze some connections between music and our body movements.

From spring 2020 on, most of the concert houses and theaters had to be closed because of the covid-19 virus. During such difficult times many musicians organize spontaneously solo, duo or trio performances, go out on the streets or into the gardens of old people's homes and play music. In this way, music gives us some bright spots in the dark world.

Music has always existed together with the life of human beings. We hope that both music and human beings can survive for a long time from now on, as well.

Dortmund, Germany   Kinko Tsuji
November 2020   Stefan C. Müller

Acknowledgments The authors are happy that Manuel Covarrubias supplied an authentic picture symbolizing the symbiosis of physics and music. The authors thank Volker Bley, Dortmund, to have offered us to take pictures of his abundant collection of historical instruments. We convey our great appreciation to Angela Lahee of Springer Verlag for efficiently supporting this manuscript. We also acknowledge Stephen Lyle for his advise about English musical terms, and Sho Tsuji for critical reading.

Contents

Part I
Physics + Music = ...

Musik ist
ein Teil des schwingenden Weltalls.

Music is
a part of the vibrating cosmos.

—*Ferruccio Busoni*

1

Introduction

1.1 Physics & Music - Quiz of the Year

Let us start this book with the following quiz. Can you identify the persons in Fig. 1.1?

If you have answered with the correct names of all six persons, you are an expert in the hybrid field of "Physics & Music". Congratulations! If you can further say what the mathematical/physical symbols signify, maybe you do not need to read this book at all. (Answers are given at the end of this chapter.)

1.2 What Is the Hybrid Field of Physics & Music?

1.2.1 Human Activities with Tones and Movements

Many animals and early humans have created tones and movements for communication: with the practical purpose to live and to survive. They must have sent signals for imminent dangers, wishes for mating, or just expression of certain moods connected with the search of food.

About 100,000–70,000 years ago the structure of the human larynx has gradually changed so as to produce vowels, leading to the ability to speak [1, 2]. In parallel, the size of the human brain increased, and humans became able to develop tools, make fire and create language [3, 4].

In the Swabian Jura region of Southern Germany an ice age musical instrument of about 40,000 years old was found [5]. This is a flute made from a wing

© Springer Nature Switzerland AG 2021
K. Tsuji et al., *Physics and Music*,
https://doi.org/10.1007/978-3-030-68676-5_1

Fig. 1.1 Who are they? (picture by Manuel Covarrubias)

bone of a griffon vulture, and there are some hand-born finger holes, as shown in Fig. 1.2. It suggests that this flute created tones with different pitches and that our ancestors of 40,000 years ago played some melodies with this flute.

Similar to the animal world, for our ancestors both tones and movements had practically the same purpose: transmission of important information for survival. Over an enormously long period of time tones and movements have been partially losing their original purposes, and have turned into more complicated forms (languages, music and dances), possessing a more ornamental characters and thus moving into the category "Arts".

On the other hand, tones and movements are issues of natural science and technology. Scientific knowledges and technologies can contribute to developing arts further. Physiologists have tried and are trying to find out, how we hear and how we move. Physicists have studied and are studying, how sounds and movements are created. Mathematicians have tried and are trying to find mathematical order and beauty in arts. Developments of mechanics and materials contribute to construct better musical instruments, as well as concert halls. Modern electronics make it possible to conserve musical and dance performance in digital forms. Such relationships are schematically shown in Fig. 1.3.

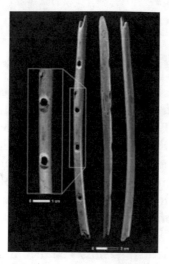

Fig. 1.2 An ice age musical instrument (about 40,000 years old) found in a cave in the Swabian Jura region in 2008. 4.2 cm length. ©H. Jensen/Universität Tübingen

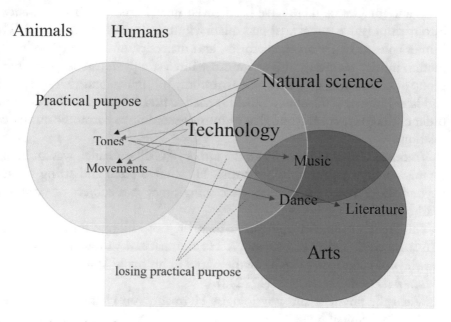

Fig. 1.3 Relationship of arts and sciences concerning tones and movements of human activities

1.2.2   Arts or Natural Sciences?

Does music belong to the arts? The answer is "Yes".

Does music belong to the natural sciences? The answer is "?", "Maybe yes", "Maybe no" or, "I do not know". Probably the answer depends on which kind of music one thinks of. If one thinks about "Das Wohltemperierte Klavier (The Well-Tempered Clavier) BWV 846–893" of Johann Sebastian Bach, the answer is certainly "Yes". If one thinks about "Erlkönig" (The Erlking) Op. 1, D 328 of Franz Schubert, the answer could be "No". Bach's work is systematic with a pronounced internal structure, while in the case of the Erlkönig a dramatic story comes together with music. If one listens to the "Concierto de Aranjuez" of Joaquín Rodrigo, emotion develops in the first place, and the answer is "?... I cannot say anything".

In our history some natural scientists studied music as a scientific object, and some musicians explained music within the field of natural science. Below we show some of these cases.

In ancient Greece there have been seven liberal arts, which were divided into trivium (three verbal arts) and quadrivium (four mathematical arts). The former consists of grammar, rhetorics, and dialectics, and the latter of arithmetics, music, geometry, and astronomy. Interestingly, music is in the same group as arithmetics, geometry and astronomy [6]. Pythagoras (570–510 BC) and later Ptolemy (100–about 170 AD) tried to find a connection between the order of the entire cosmos and that of music according to harmonic numerical relations.

Vincenzo Galilei (1520–1591), the father of Galileo Galilei, was an Italian lutenist, composer, and music theorist. He studied pitch and string tension beyond the Pythagorean tuning, describing them as a subject of nonlinear mathematics [7].

The French composer and music theorist Jean-Philippe Rameau (1683–1764) is called "father of harmony". He summarized various intervals with drawings of coordinates like in mathematics in his book *"Traité de l'harmonie réduite à ses principes naturels"* (1722) [8].

When the physicist and physiologist Hermann von Helmholtz published in 1877 his seminal book *"Die Lehre von den Tonempfindungen als physiologische Grundlage für die Theorie der Musik"* [9] he wrote in the first lines of his introduction: "In the present work an attempt will be made to form connection between the boundaries of two sciences, which, although drawn together by many natural relations, have hitherto remained sufficiently distinct—the

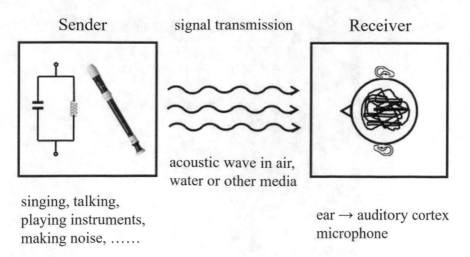

Fig. 1.4 Signal transmission from sender to receiver

boundaries of physical and physiological acoustics on the one side, and of musical science and esthetics on the other. The class of readers addressed will, consequently, have had very different cultivation, and will be affected by very different interests."

We could "buy" this text for our own introduction to this book.

1.2.3 Sender and Receiver

Music should be played and what is played should be heard. Theoretically, this process can be explained by the signal processing in physics (see Fig. 1.4).

The source of the sound/noise (Sender in Fig. 1.4) can be anything: not only music instruments played in a European orchestra but also grass flutes, drums, rattles, and many instruments from non-European countries, such as the Jew's harp (see Fig. 1.5).

The most frequent principle of sound transmission is the following:

A sound source is excited by a linear or surface vibration. A resonance body which is attached to the sound source amplifies the vibration, radiating into the surrounding air. Human ears function as a receiver. The acoustic waves reach our ears and are transmitted into the auditory cortex.

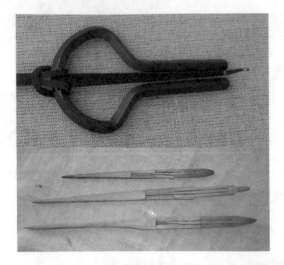

Fig. 1.5 Jew's harp is an instrument consisting of a flexible metal or bamboo tongue or reed attached to a frame. Top: Slovakian jew's harp "drumb'a" (CC BY 4.0 by Krzysiu), bottom: Assam jew's harp "gogona" made from bamboo (CC BY-SA 3.0 by Abhi kal)

1.3 Early History of Musical Notations

There are certainly many people who have not been familiarized to our system of musical notations, and the corresponding scripts are as difficult to decipher as a Chinese text. One has to learn musical notations like a new script from a different language. Many have done such things readily during their education, like learning the characters of our alphabet.

It took thousands of years to develop our alphabets to their current form. Just consider the character A: in the Phoenician alphabet (around 1000 BC) the first character was aleph (= bull head) symbolized as in the sketch in Fig. 1.6, top. This character was turned around and later took the form of the Greek Alpha A or the Latin A. Similarly, the third letter gaml symbolized by the camel's back and turned into the Greek letter Gamma Γ, precursor of our C or G (Fig. 1.6, bottom).

1.3.1 Mesopotamia, Egypt and Greece

For the development of musical notation we also look back on many centuries. Melodies have certainly been handed down by listening and copying in early times (even now in some areas). There must have been at a point in time the

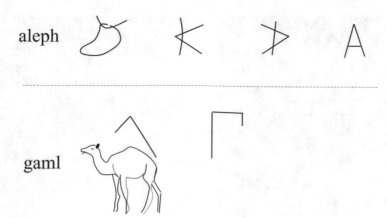

Fig. 1.6 Development of our alphabet. Top: from aleph to A, bottom: from gaml to Γ

need to find a way to write them down for further use. We have an impressive example from the times of ancient Mesopotamia.

In the early 1950s, archaeologists unearthed several clay tablets from the 14th century BC in the ancient Syrian city of Ugarit. As shown in Fig. 1.7, these tablets contained cuneiform signs[1] in the Hurrian language, which turned out to be the oldest known piece of music ever discovered, a 3400 year-old cult hymn [10].

Egyptian and other ancient cultures must have had some kind of musical notation for documentation, but the first completely deciphered one was

Fig. 1.7 Old music notes of Hurrian hymn of the 14th century BC in the ancient Syrian city of Ugarit (public domain, author unknown)

---

[1]Cuneiform is a writing system with "wedge-shaped" marks. It is one of the earliest systems of writing, invented by Sumerians in ancient Mesopotamia.

Fig. 1.8 Left: the Seikilos column made with marble, engraving poetry and musical notation, right: the inscription in detail. Nationalmuseet Denmark (CC BY-SA 3.0 by Nationalmuseet)

created by the Greek (Seikilos epitaph, Fig. 1.8), perhaps already around the 3rd century BC. Letters and other signs were used for pitch and tone duration—supposedly derived from the first 7 characters in their alphabet, corresponding to the strings of the kithara. (A kithara is exhibited in the gallery of historical instruments, below in Sect. 1.4.) Other symbols written above the line served to indicate the tone duration [11].

1.3.2 Notation Systems in the Middle Ages

In Europe this symbolism got lost with the end of the Roman Empire and only little was conserved and later deciphered. Outside Europe, though, there evolved different notation systems, in China as well as in India. Since the 13th century the Arab notations became known, based on the handed down Greek tradition. Quite generally, the music notation served then to memorize improvised music and less to conserve melodies for generations to come.

In the Middle Ages (from the 9th century AD on) European monasteries started a new type of notation for Gregorian chorals: the **neumes** - graphical symbols to identify certain melodic phrases, as shown in Fig. 1.9 [12]. The neumes serve to indicate by a kind of stenographic arrows the approximate

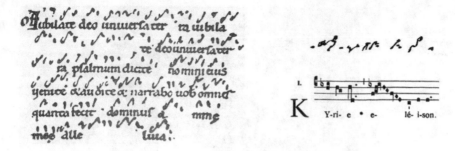

Fig. 1.9 Left: cheironomic neumes (neumes without lines) describing psalm verses "Jubilare deo universa terra", right bottom: neumes with 4 lines, a sample of Kýrie Eléison XI (Orbis Factor) from the Liber Usualis; some times the cheironomic neumes appear on top of the lines (public domain)

course of a melody. Lines were added, as well, starting with just two for tones F and C.

It was then Guido da Arezzo (991/992—after 1033),[2] who at the beginning of the 11th century added parallel lines to the already existing lines for notes F and C in order to grasp precisely the tone intervals between the lines. He also invented the clef, at the time mainly in terms of the C-clef.

Guido da Arezzo is also known for the Guidonian hand (Fig. 1.10). This was a mnemonic device for sight-singing. (This kind of device had existed before his time, though.) He used the joints of the hand for teaching his hexachord and solfège (a method of naming pitches by using syllables) [13].

Since the 14th century **tablatures** have become common, where the number of lines corresponds to the number of strings on which the position of the gripping hand is marked, as depicted in Fig. 1.11 [14]. This approach has survived until today, e.g., with chord notations for the guitar.

Whereas Guido da Arezzo's invention worked with 3 and later 4 lines (still in use in some of church music), we are now accustomed to a 5-line system, which came into use in France in the 16th century and developed to be the basis of modern music writing. Five lines proved to be well adapted to the average capacity of the human voice. This modern system will be used throughout the following chapters.

---

[2]His exact years of birth and death are unclear. The year of death could be around 1050.

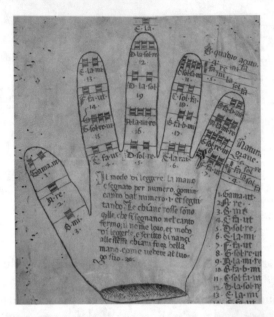

Fig. 1.10 Guidonian hand from a manuscript from Mantua, last quarter of 15th century (Oxford University MS Canon. Liturg. 216. f.168 recto) (Bodleian Library) (NB: Color image is unsourced–black-and-white image in file history from MS Canon Liturg. 216) (public domain)

Fig. 1.11 Tablatur notes: Fol. 44r from "Robertsbridge Codex" with the beginning of "Tribum, quem non abhorruit". British Museum (public domain)

1.4   Gallery of Historical Instruments

In early times many people on the continents of Africa, America, Asia and Europe have developed their own ways to sing and play pentatonic melodies. These reflect their relation to folk music and sounds for other common needs and events. We describe non-European music in Chap. 5. However, in this section we introduce some musical instruments which show a few quite remarkable cultures, where the pentatonic or a similar musical system has been cultivated.

Fig. 1.12 The woman standing to the right plays a harp: The three musicians, Tomb of Nakht, Thebes (public domain from The Yorck Project (2002) [15])

Lyres and harps count among the oldest instruments with several strings, originating in Sumerian and Mesopotamian culture (see Fig. 1.12). From Egypt they found their way southwards to Nubia and East Africa, including Ethiopia.

If we have a look at Ethiopia, Eritrea and Somalia, we detect a historically popular instrument, the krar. As seen in Fig. 1.13 (left), this plucked instrument, belonging to the class of lyres, has 5 or 6 cords guided in a trapez-like manner across a resonating corpus made of wood. The strings are tuned along a pentatonic scale which can be adapted to the voice pitch of the singer [16]. The instrument to the right in Fig. 1.13 is a pluriarc of West Africa. Five strings are spanned from the neck to the bridge. It is played by plucking the strings either with fingers or with a plectrum [17].

In Greece, where so many rules on tones and scales were invented, a traditional instrument was also made of strings mounted on a wooden frame: the kithara (ancient root for the word "guitar"), played mainly for noble events, in

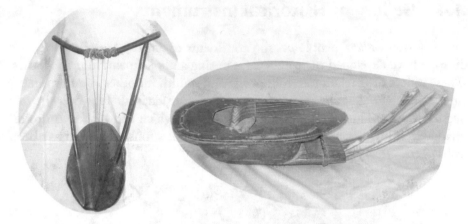

Fig. 1.13 Left: krar, right: pluriarc (Photos by KT in Volker Bley's gallery)

particular in honor of the god Apollo (see Fig. 1.14). In parallel one used the lyre, a smaller instrument without a foot basis [18].

Fig. 1.14 Muse tuning two kitharas. Detail of the interior from an Attic white-ground cup from Eretria, in 465 (photo by Jastrow (2005): public domain)

The kithara had 5–12 strings and came about in the 8th/7th century BC. For playing see the person standing on the right in Fig. 1.15. The player fixed the instrument on his breast. His right hand plucked the strings with a plectrum, while the left hand damped the sound and provided higher tones by shortening the strings [19].

Wind instruments were widely used for playing pentatonic music, with flutes and reed instruments being the most popular among them. The aulos was an exotic double reed woodwind instrument, a precursor of the modern oboe. It existed already in ancient Egypt and Italy. The primary form consisted of a pair of cylindrical (conical) bore pipes with double reed mouthpieces and finger holes (see the person in the middle of Fig. 1.15). Early aulos were made from reed stems, wood or bone.

Fig. 1.15 Dancers and musicians, tomb of the leopards, Monterozzi necropolis, Tarquinia, Italy. UNESCO World Heritage Site. Fresco a secco. Height (of the wall): 1.70 m. (public domain)

We move to the north: The Finnish kantele or Estonian kannel (Fig. 1.16) is a box zither with 5 pentatonic strings fixed on a burnt-out and carved-out birch trunk, frequently tuned in D major or D minor. Later on the string number was increased up to as many as 23 strings. It can be played in two ways: either the musician puts the long strings directly at his front (Haapavesi style), or the more traditional player would turn the shorter strings towards himself (Perhonjoki-style) [20].

Fig. 1.16 Top: Finnish kantele with 5 strings (CC BY-SA 3.0 by TheYellowFellow), bottom: Estonian kannel with 6 strings (CC BY-SA 3.0 by Adeliine)

Fig. 1.17 Left: a Zen Buddhism monk blows shakuhachi in walking (CC BY-SA 3.0 by Corpse Reviver), right: nose flute played by a Paiwan (The Paiwan are an indigenous people of Taiwan) (CC BY 2.0 Presidental Office Taiwan)

Moving on to Asia, China founded a tonal system similar to the one that Pythagoras achieved. However, later it was developed in ways different from the European one. The shakuhachi is an ancient Chinese longitudinal, end-blown bamboo flute (chĭ bā), which was introduced to Japan in the 7th century. The instrument is tuned to the minor pentatonic scale. As shown in Fig. 1.17 left, Zen Buddhism monks used this instrument for "blowing meditation" [21]. There are 4 holes on the front and one hole on the back side.

A nose flute (Fig. 1.17 right) is a popular musical instrument played in South East Asia, Polynesia, the Pacific Rim and in Africa. It is made from a single bamboo section and has up to four finger holes. There are a recorder type as shown here and a transverse type.

Its provenance is the Persian area and, around the year 1500, it was transmitted through Islam to the Asian East including Malaysia and Thailand. Its tonal range is two octaves. Due to its thin or sharp penetrating sound, it is preferentially used in large orchestras. Similar instruments have been in use in countries like Northern India (tangmuri), Tibet (gyaling) or Nepal (mvali).

The Kulintang is a horizontal row of "nipple gongs" played on the south-Philippine island Mindanao [22]. The instrument consists of a row/set of 5 to 9 graduated pot gongs, horizontally laid upon a frame arranged in order of pitch with the lowest gong found on the player's left, based on pentatonic principles (see 8 gongs on the table in Fig. 1.18).

Fig. 1.18 A set of gongs of Kulintang on a table, Insets are a top view of the two gongs (bottom middle) and sticks (bottom right) (CC BY-SA 2.5 by Philip Dominguez Mercurio)

The saung (a Burmese string instrument) is an arched horizontal harp and is said to be the only surviving harp in Asia [23]. Different from the European harps, the resonator is placed horizontally, as shown in Fig. 1.19. There are thirteen or sixteen strings, which have been traditionally made of silk (now often made of nylon).

Fig. 1.19 Saung musician in 1900 (public domain)

The Indonesian islands of Java and Bali are home to the gamelan music. It consists of many gongs, sounding plates, some xylophones, drums, and perhaps flutes, strings or voices. Noteworthy is a relief panel on a wall of Borobudur: a musical ensemble, probably a gamelan ensemble is engraved (Fig. 1.20). Borobudur is the world's largest Buddhist temple built in the 9th century in Central Java, Indonesia [24].

Finally we arrive on the American continent. The wall paintings dating from 775 AD at the Bonampak ceremonial complex [25] near today's border of Guatemala (Fig. 1.21) indicate that in the Maya period people played instruments such as trumpets, flutes, whistles, and drums.

A horizontal log drum, the "teponaztli", played an important role in the Aztecs culture (see Fig. 1.22) [26].

Fig. 1.20 Musicians performing in a musical ensemble, bas-relief of Borobudur, probably an early form of gamelan (CC BY 3.0 by Gunawan Kartapranata)

Fig. 1.21 Maya musicians: twin trumpeters standing side by side in a 12(?)-man orchestra, blowing higher overones; in Bonampak temple (public domain, author unknown)

Fig. 1.22 Left top: Two teponaztli, Aztec, found in Colima with a length of 59 cm (exhibited in the American Museum of Natural History, New York) (CC BY-SA 3.0 by Madman2001), left bottom: section through a drum, right: a drawing from the 16th century Florentine Codex showing a One Flower ceremony with two percussion instruments, a teponaztli (foreground) and a huehuetl (a big drum in the background) (public domain)

Appendix: Answers to the Physics & Music Quiz

This is a snapshot of a concert of a piano quartet with additional bongos.

Piano: Werner Heisenberg
1st violin: Ludwig van Beethoven
2nd violin: Albert Einstein
Cello: Jacqueline du Pré
Viola: Antonín Dvořák
Percussion: Richard Feynman

Equation above the piano: Heisenberg's uncertainty principle [27]

$$\Delta x \, \Delta p \geq \hbar$$

where Δx is the standard deviation of position, Δp the standard deviation of momentum and $\hbar = h/(2\pi)$ the reduced Planck constant.

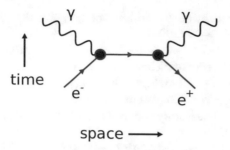

Fig. 1.23 Feynman diagram of electron/positron annihilation. Details see the review article of Kaiser [29]

Mathematical symbol between Einstein and du Pré: Einstein's field equation (general relativity) [28]

$$\frac{8\pi G}{c^4} T_{\mu\nu} = G_{\mu\nu} + \Lambda g_{\mu\nu}$$

where $T_{\mu\nu}$ is the Einstein tensor, Λ the cosmological constant, $g_{\mu\nu}$ the metric tensor, c the speed of light in vacuum and G the gravitational constant.

Diagram between Feynman and Heisenberg: a Feynman diagram (Fig. 1.23).

References

1. J. Nichols, The origin and dispersal of languages: Linguistic evidence, in *The Origin and Diversification of Language*, ed. by N. Jablonski, L.C. Aiello (California Academy of Sciences, San Francisco, 1998), pp. 127–170
2. C. Perreault, S. Mathew, Dating the origin of language using phonemic diversity. PLoS One **7**, e35289 (2012)
3. D. Bickerton, *Adam's Tongue: How Humans Made Language, How Language Made Human* (Farrar, Sraus & Giroux, New York, 2009)
4. N.D. Cook, *Harmony, Perspective and Triadic Cognition* (Cambridge University Press, Cambridge, 2012), pp. 216–219
5. N.J. Conard, M. Malina, S. Münzel, New flutes document the earliest musical tradition in southwestern Germany. Nature **460**, 737 (2009)
6. E.B. Castle, *Ancient Education and Today* (Penguin Books, Baltimore, 1969)
7. H.F. Cohen, *Quantifying Music: The Science of Music* (Springer Netherlands, Dordrecht, 1984), pp. 78–84
8. J.-P. Rameau, *Traité de l'Harmonie Réduite à ses Principes Naturels* (Dr. Jean-Baptiste. Christophe Ballard, Paris, 1722)
9. H. Helmholtz, *Die Lehre von des tonempfindungen, als physiologische Grundlage für die Theorie der Musik* (Friedrich Vieweg und Sohn, Braunschweig, 1863)

10. M.L. West, The Babylonian musical notation and the Hurrian melodic texts. Music Lett. **75**, 161–79 (1994)

11. T.J. Mathiesen, *Apollo's Lyre: Greek Music and Music Theory in Antiquity and the Middle Ages* (University of Nebraska Press, Lincoln, 1999)

12. D.G. Sunol, G.M. Durnford, *Textbook of Gregorian Chant According to the Solesmes Method* (Kessinger Pub. Co., Whitefish, 2003)

13. W.G. McNaught, The history and uses of the solfège syllables. PMRA **19**, 35–51 (1893)

14. J. Caldwell, *English Keyboard Music Before the Nineteenth Century* (Dover Publication, Mineola, 1985)

15. 10.000 Meisterwerke der Malerei (DVD-ROM), distributed by DIRECT-MEDIA Publishing GmbH. ISBN: 3936122202

16. A. Kebede, The bowl-lyre of Northeast Africa. Krar: the devil's instrument. Ethnomusicology **21**, 379–395 (1977)

17. U. Wegner, *Afrikanische Saiteninstrumente* (Museum für Völkerkunde, Berlin, 1984), pp. 82–92, 153f

18. J.G. Landels, *Music in Ancient Greece and Rome* (Routledge, London, 1999)

19. O.J. Brendel, *Etruscan Art* (Yale University Press, New Haven, 1995)

20. R. Apanavioius, *Ancient Lithuanian Kanklės* (Institute of Ethnomusic, Vilnius, 1996)

21. J, Keister, The Shakuhachi as a spiritual tool: a Japanese buddhist instrument in the west. Asian Music **35**, 104–105 (2004)

22. K. Benitez, *The Maguindanaon Kulintang: Musical Innovation, Transformation and the Concept of Binalig* (University of Michigan, Ann Harbor, 2005)

23. T. Miller, S. Williams, *The Garland Handbook of Southeast Asian Music* (Routledge, London, 2008)

24. https://en.wikipedia.org/wiki/Borobudur#cite_ref-Raffles1814_13-1

25. M.D. Coe, *The Maya* (Thames & Hudson, London, 1999)

26. M.D. Coe, *Mexico: From the Olmecs to the Aztecs* (Thames & Hudson, London, 2002)

27. W. Heisenberg, Über den anschaulichen Inhalt der quantentheoretischen Kinematik und Mechanik. Z. Phys. **43**, 172–198 (1927)

28. A. Einstein, *Relativity: The Special and General Theory, translated in 1920* (H. Holt and Company, New York, 1961)

29. D. Kaiser, Physics and Feynman's diagrams. Am. Sci. **93**, 156–165 (2005)

Part II
Our Modern Tonal System

Music is a science which should have definite rules;
these rules should be drawn from an evident principle;
and this principle cannot really be known to us
without the aid of mathematics

—*Jean-Philippe Rameau*

2

Notation and Tonal Systems

2.1 What Is .....?

The notion of art refers to creativity found in human societies, both in theory and in physical expression. In history there have been many ways for classifying art depending on the cultural environments and viewing some fields as the main arts, others as derivatives of them. In modern times major constituents include literature (drama, poetry, prose), performing arts (among them dance, music, and theatre) and visual arts (comprising painting, sculpting, architecture, filmmaking, and others).

2.1.1 Music

We will talk about music first.

Music is an art form and cultural activity whose medium is sound organized in time. It gathers its material from the realm of tones. It thus belongs to the category of performing arts, different from the visual and/or fine arts, which are mainly static in nature.

The common elements which characterize any piece of music are melody, rhythm and harmony. Melody and rhythm proceed in time, while harmony acts at any point of the proceeding line of melody. Later we will devote more time on these characteristics (in Chap. 3), but will first mention a few other important properties of music.

© Springer Nature Switzerland AG 2021
K. Tsuji et al., *Physics and Music*,
https://doi.org/10.1007/978-3-030-68676-5_2

2.1.2 Sound

Sound is a complex system formed from single tones: a basic tone plus several resonating single tones give rise to a series of overtones as analyzed in acoustic spectra. This creates the sonic qualities of timbre of a musical sound resulting in an intuitive tone color: glaring, soft, full, muffled, dark, dull, etc. Much more about that in Chap. 4.

2.1.3 Tone

The tone is a sound signal, as generated by periodic vibrations of an elastic body. Its pitch (tone height) is determined by vibrations of the air measured in Hertz ($1\,Hz = 1/s$) for frequency.

Our perception tells us that low frequency tends to sound heavy, dull, earthy and that high frequency sounds light, floating, piercing.

The intensity (loudness) of the tone is determined by the amplitude of vibrations, the dynamics of a melody by variations of amplitude. Its sound has a color as described in Sect. 4.4. Its duration is indicated by a system of different symbols in connection with rules for "tempo".

2.1.4 Noise

If a sound has a continuous frequency spectrum comprising many kinds of sounds emitted from different sources, we call it noise. That could be: the winds and waves, usual background in daily life, percussion instruments, explosions, ... Many types of noise have band-limited spectra—one hears a dominant tone (as with timpani) or they are "white" not permitting to perceive any tonality (rattles).

2.1.5 Spectral Characteristics of Sounds

Sounds and tones, the elements of music, have some basic properties, which we will show now in a qualitative way. A tone has a well-defined pitch, that can be measured by its frequency. We mean frequencies between 20 Hz and 20,000 Hz, which can be heard with our ears [1]. As a useful characterization of sound, that has a color, we need to determine a whole spectrum of frequencies representing all the overtones (basic tone plus series of harmonics - see Chap. 3).

In Fig. 2.1 we assemble a qualitative list of tones, sounds and their frequency spectra.

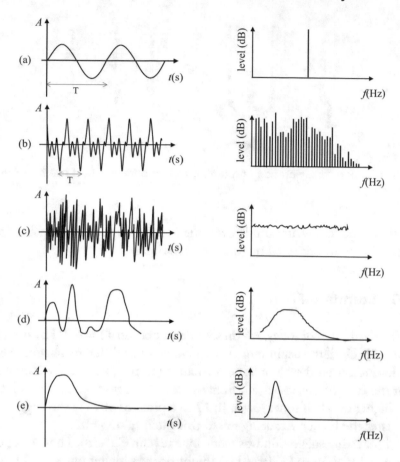

Fig. 2.1 Time courses (left column) and frequency spectra (right column): **a** sinusoidal sound and its single line spectrum; **b** sound and multiple overtone spectrum; **c** "white" noise and atonal spectrum; **d** band-limited noise and continuous (tonal) spectrum; **e** bang and its broad, line-shaped spectrum

Example: The Tuning Fork

It is not so easy to produce a pure sinusoidal sound with any musical instrument, there will be always some overtones of the fundamental frequency involved.

It is the tuning fork, (invented in 1711 by the trumpeter John Shore, see Fig. 2.2a, which emits a dominantly sinusoidal signal (as shown in Fig. 2.1a), adjusted to the standard pitch A4 = 440 Hz, as defined in 1939 as a reference tone for the musical instruments of an orchestra [2]. Beyond the basic tone there are just very weak harmonics due to different vibrations of the fork as shown in Fig. 2.2b, c and d. Here we find sine sound vibrations (with very small

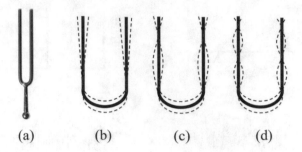

(a)   (b)   (c)   (d)

Fig. 2.2 a Tuning fork; vibration of a tuning fork: **b** pure sine, **c, d** overlay of several harmonics

deviations due to low amplitude overtones) and a more complex vibration, if the harmonics are activated by (too) strong stimulation.

2.1.6 Standard Pitch

The standard pitch of today, A4 (in German notation a') $= 440$ Hz, has a long history [3]. Until the beginning of the 19th century different versions were in use, depending on the place of performance (church, opera, chamber music, ...) or the type of music played. Renaissance music used pitches up to 4 half tones higher or 3 half tones lower. In 1788 a pitch of 409 Hz was suggested in Paris, later the French Academy raised this pitch to 435 Hz.

An interesting suggestion was made by Sauveur/Chladni. They set a period of 1 s equal to a "deep C" ($= 1$ Hz). Eight octaves higher one would find the C4 with a frequency of $2^8 = 256$ Hz, leading to A4 $= 426.666...$ Hz (in just intonation).

Nowadays there are several norms for the standard pitch issued by standard organizations of many countries. These regulations are taken with a "maybe" or "must" attitude. In Austrian and German orchestras one has introduced 443 Hz for standard pitch, because then the string instruments sound louder and fuller. In other countries the standard pitch may vary between 440 and 445 Hz.

The tone with standard pitch is played, before a rehearsal starts, by the oboe, then taken over by the concert master. His playing this note is the signal for all other members of the orchestra to tune their instruments correctly.

As mentioned above, there does not exist a unique determination of the standard pitch. Soprano singers have complained that this tone height should be reduced to be adapted to the capacities of their voices.

Different standard pitches are not appreciated by woodwind players: Whereas string instruments can be easily tuned somewhat higher or lower, there is less margin for woodwinds to be adjusted to different pitch.

2.2 What Is Physics, in Close Partnership to Music?

When Dr. Faust, in his first monologue, contemplates all what he has studied throughout his life, he is dissatisfied with his understanding of how the world works and disappointed that science has failed him in his research into "what holds the world together at its very core" (Goethe, Faust I).

The physicist may believe that this touches a lot his domain of responsibility. This may be true to some extent. However, many other scientific disciplines would have to come into play. In the framework of this book the most relevant subdivisions of physics are mechanics, acoustics (fundamental and technical), thermodynamics of gases, vibrations and waves, nonlinear dynamics. In addition, scientists from material science, signal processing, music theory and mathematics, physiology and neurobiology will raise their voice, when we arrive at sections of their expertise. Thus we are facing, in fact, an interdisciplinary endeavor.

To get started we refer to a historic contribution by one of the great pioneers in the field of our interest: Hermann von Helmholtz (1821–1894), German physicist and physiologist, who published in 1863 the seminal book *"Die Lehre von den Tonempfindungen als physiologische Grundlage für die Theorie der Musik"* [4]. This work comprises a multitude of illuminating experiments and useful ideas at the borderline between acoustics and physiology.

Among well-known inventions we find, for instance, the Helmholtz resonator shown in Fig. 2.3. It consists of an air volume of arbitrary shape (usually a sphere) with a cylindrical small neck (indicated by "a" in Fig. 2.3) and an opening to the outside "b". Such a resonator can be quite small. The thin end "b" is carefully introduced into the ear canal, where it acoustically couples to the eardrum and allows for detailed listening. A useful purpose is to separate a single fundamental note from a mixture of sounds of different frequencies. The air volume inside the sphere has a mass and elasticity with a pronounced eigenfrequency, which can be heard much better than all other possible modes. An efficient application exists for deep tones of musical instruments, like in a bass marimba, contrabassoon, or bass clarinet.

On the other hand, in acoustics one uses also very large "laboratories", like concert halls, in which the room acoustics has to be characterized and then

Fig. 16a.

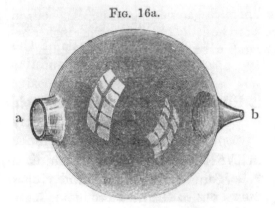

Fig. 2.3 Helmholtz resonator, Fig. 16a in his book [4] (public domain)

Fig. 2.4 Measurement of a diffuser in an anechoic chamber (CC BY-SA 3.0 by Trevor Cox)

optimized. For the study of room acoustics one frequently works in rooms of low reflection, where the walls and ceiling as well as the floor are covered with absorbers, mostly wedge-shaped or lamella, made from some absorbing material. Such a room is called an anechoic chamber (see Fig. 2.4). The object of study is placed somewhere close to the center.

Interestingly, in an anechoic chamber the principle of the Helmholtz resonators may serve as that of resonance absorbers: there, impinging acoustic energy (the kinetic energy of the mass) is transformed to heat. Maximum absorption occurs in the range of the eigenfrequency, where the mass vibrates with highest amplitude. Therefore, the Helmholtz approach is applied mainly for low-frequency spatial modes in concert halls, theaters, studios, office-space or conference rooms.

In fact, acoustical research is conducted in musical systems with small and large spatial scales. This is obvious in the realm of musical instruments: start with the Jew's harp (see Fig. 1.5) at the lower end of sizes and move on to the organ of a cathedral.

Considering music in the context of natural science, we will take the opportunity to present and discuss, whenever it becomes necessary in the course of the chapters to follow, cross-links appropriate to define, describe and analyze relevant scientific and mathematical relationships and properties.

2.3 Tonal System

2.3.1 Lines and Notes

We start from five horizontal, equidistant lines. There could be less or more lines, but since the 16th century a number of five has been generally accepted as a useful standard. (Four lines are still common in Gregorian melodies and some church chorals, as described in Sect. 1.3.2). Authors wonder, why this five-line system has been kept unchanged for more than 500 hundred years, although it is quite irrational. Why is the tone interval of the two adjacent lines not always the same?

In the 5-line system, also called the staff (US), the stave (UK), or staves (plural for both), one can represent 11 different musical pitches and various intervals. Figure 2.5 shows (a) some examples of intervals (details in Sect. 3.1), (b) elements of a note, (c) notes with various durations (each note in this sequence has half the length of the previous one) and (d) ledger lines which are used when tones are lower or higher than the 11 pitches in the staff.

For adjacent notes the flags are mostly written as

The left figure shows two adjacent quarter notes, the middle one two sixteenth notes and the last group is called "triolet" (three adjacent thirds of the quarter).

There are also dotted notes for intermediate length. For example, for the length of a 3/4 note (between a whole note and a half note), one adds a dot to

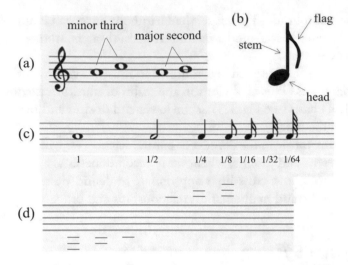

Fig. 2.5 a Examples of intervals, **b** elements of a note, **c** notes with different length 1 = whole note, 1/2 = half note, 1/4 = quarter note (crochet), 1/8 = eighth note (quaver), 1/16 = sixteenth note (semiquaver)..., **d** ledger lines

a half note; for the length of a 3/8 note one adds a dot to a quarter note; and for the length of a 7/16 one adds two dots to a quarter note as shown below.

2.3.2 Rest

For a rhythmically satisfying performance of a musical piece the precise timing of rests (a musical notation that indicates the absence of a sound) is as important as that of the notes. Accordingly, one has introduced symbols for rests of any desired length.

2.3.3 Clef

A clef at the beginning of a staff determines the pitch for a given single line (out of 5). An evolution of the form of clefs has taken place during the centuries. When church modes were dominating, around the time of Guido da Arezzo, who invented the line system for musical notations, after early attempts the design of the C-clef emerged.

It focuses in its center on the middle line for C4 (see Sect. 2.3.5). Shifting this key up or down alters the assignment of the actual C4 note.

development of the C-clef (from left to right)

Around the same time the F-clef came into use, pointing at the F3 and derived from the capital letter F as shown below. When higher pitch became common, the G-clef was introduced, mainly applied to violin or flute music.

development of the F-clef (top) and G-clef (bottom)

2.3.4 Bars

In the musical scores there appear vertical lines, called "bar lines" as shown in Fig. 2.6. A standard bar line (or measure) indicates a segment of time defined by a given number of beats. A double bar line (or double bar) separates two sections within a piece. For the ending of a piece (or a movement) a double bar line with a combination of a standard line and a bold line is used. A pair of repeat bar lines (for starting and ending, Fig. 2.6, middle) indicates that the passages between these bar lines should be repeated. If the left repeat bar line is absent (Fig. 2.6, right), it means that the repeat should start from the beginning of the piece (or movement).

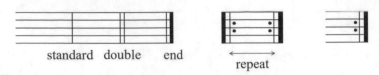

Fig. 2.6 Various bar lines

Normally the first beat of a bar is emphasized by a leading musician or the conductor. The latter would usually mark the beat with a long stick. The organization of notes within a bar is important for the structuration of a melody, of a motive and for the rhythmic evolution of a musical piece.

> ### *Intermezzo*
>
> Sadly, the French composer Jean-Baptiste Lully (1632–1687) stabbed his toe with a long stick, when he was conducting a Te Deum to celebrate Louis XIV of France's recent recovery of Louis XIV of France from illness in 1687. The banging was so strong that it created an abscess. The infection was spread rapidly, and he deceased a few weeks after.

Number and type of notes or beats for each bar are indicated by a time signature (a pair of numbers at the start of a musical piece: a broken rational consisting of numerator and denominator). For example: the upper number of **4/4** counts the number of beats in the bar, the lower number indicates the note value of each of the beats, having 4 beats of the length of a quarter note in each bar. With **3/4** the note value is the same, but only 3 beats are required for each bar.

Correspondingly one would interprete **9/8** as 9 beats with eighth length to be played.

The **4/4** bar can be described also with a special symbol like a "C". If a vertical bar is drawn through "C", this signifies the **2/2** bar: a half note is required as basic beat (alla breve), which signals a faster tempo.

$$\mathbf{\frac{4}{4}} = \mathbf{C}, \quad \mathbf{\frac{2}{2}} = \mathbf{\mathcal{C}}$$

Most common types of beats are (beat separation by vertical lines):

The beats marked with ">" mean "strong". For example, for the **2/2** as well as the **2/4** one plays usually: <strong, weak>, <strong, weak>..., for the **3/4** and the **3/8**: <strong, weak, weak>, <strong, weak, weak>..., and for the **4/4**: <strong, weak, intermediately strong, weak>, <strong, weak, intermediately strong, weak>,...

The upbeat is an incomplete bar at the beginning of a melody or a musical subunit. It consists of one or a few more notes without special emphasis played just before the regular timing of the beats begins, as shown below for the beginning part of "Auf Flügeln des Gesanges (On wings of song)" of Felix Mendelssohn Bartholdy on a poem of Heinrich Heine.

upbeat

Beginning of "Auf Flügeln des Gesanges (On wings of song)" of Felix Mendelssohn Bartholdy with upbeat

2.3.5 Pitch

In physics the pitch is determined by frequency in Hertz (1 Hz = 1/s), while in music it is characterized by letters, syllables, special characters, and specific graphical symbols.

German absolute designation of notes developed during musical history following the alphabet: The seven natural notes[1] (in German: "Stammtöne") within an octave are named by the letters:

A, H, C, D, E, F, G.

In English notation one finds:

A, B, C, D, E, F, G.

There is the "strange" difference between H and B.

---

[1] Here natural notes means the notes without accidentals. They are different from the natural tones which are created as harmonics.

Brief History

When Guido da Arezzo published his musical rules in 1025 AD, he considered the theory of hexachords* (probably derived from the six-string Greek lyre and basic for learning Gregorian singing). The so-called hexachord durum (durum meaning "hard") required in the church modes the note "B durum", whereas, due to the developing rules of church modes one had to use in the hexachord molle (molle meaning "soft") the slightly lower note "B molle". The original B durum was written as a small angular b (♭: b-quadratum), the B molle half tone lower as a round b-rotundum (♭). In the English tradition the b-quadratum became the note B, while the b-rotundum became Bb (B flat). In Germany, on the other hand, the b-rotundum became the note B (b), whereas the b-quadratum was hitherto labeled H for distinction from the B because of its similarity to h, which was also more appropriate for printing.

This signaled the start of writing the "accidentals" used for indicating, whether a note should be elevated (#) or lowered (b) by a half tone. The b-quadratum developed to the sharp # (German: Kreuz) and the natural sign ♮ (German: Auflösungszeichen), the b-rotundum stands for the flat ♭ (German b).

* For details see Chapter 3

With our 5-line system and in G-clef one writes for the C major scale:

C   D   E   F   G   A   B   C

Continuing the scale the notes upwards or downwards are repeated, the names of the notes are replicated due to octavation.

The absolute tone levels were originally named after the alphabet with the first tone A taking the role of a standard pitch (today at 440 Hz cf. Sect. 2.1.6). Later one used to start with C instead of A. Table 2.1 lists names of these tones in various languages.

Table 2.1 Names for the tones are multilingual (since the 10th century)

German	C	D	E	F	G	A	H
English	C	D	E	F	G	A	B
Italian	do	re	mi	fa	sol	la	si
French	ut/do	ré	mi	fa	sol	la	si
Russian	do	re	mi	fa	sol	la	si
Japaese	ha	ni	ho	he	to	i	ro

The piano illustrates best the structure of the whole tonal system as it is used in modern occidental music. In the following figure (Fig. 2.7) we show all the keys (88 keys) of the piano and corresponding notes. The white keys represent the natural notes.

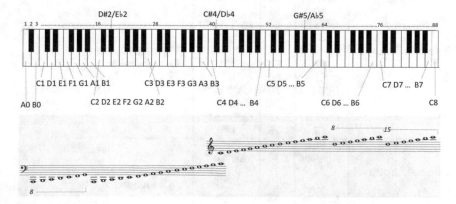

Fig. 2.7 Keyboard of a piano and corresponding names and notes of each key

In this figure tones coming from white keys are grouped in octaves of the C major key. Each octave has a specific denomination which provides a corresponding label for the tones belonging to it. Two notes A0 and B0 belong to the sub-contraoctave followed upward by the contra-octave with notes C1, D1,...to B1. Higher notes comprise the great octave, small octave, one-line octave, two-line octave, and further (see Table 2.2).

2.3.6 Change of Pitch—Accidentals

For changing the pitch of a note by a half tone we have the accidentals. A sharp (♯) in front of a note increases, a flat (♭) decreases the pitch by a half tone. This prescription is then valid within the corresponding bar. (In German, the name of the tone with ♯ is indicated by the suffix -is, and that with ♭ by the suffix -es: exceptions: as, es, and b.) The natural sign (♮) cancels a preceding sharp or flat in the same bar.

sharp   flat   natural

Table 2.2 Comparison of English and German notation

English notation	Piano key		German notation
	C8	88	c'''''	Fünfgestrichenes c
	B7	87	h''''	
	A♯7/B♭7	86	ais''''/b''''	
	A7	85	a''''	
	G♯7A♭7	84	gis''''/as''''	
	G7	83	g''''	
7(four-lineoctave)	F♯7/G♭7	82	fis''''/ges''''	Viergestrichener
	F7	81	f''''	Oktavbereich
	E7	80	e''''	
	D♯7/E♭7	79	dis''''/es''''	
	D7	78	d''''	
	C♯7/D♭7	77	cis''''/des''''	
	C7	76	c''''	
	B6	75	h'''	
	A♯6/B♭6	74	ais'''/b'''	
6(three-lineoctave)	.....	...	.....	Dreigestrichener
	.....	...	.....	Oktavbereich
	C♯6/D♭6	65	cis'''/des'''	
	C6	64	c'''	
	B5	63	h''	
	A♯5/B♭5	62	ais''/b''	
5(two-lineoctave)	.....	...	.....	Zweigestrichener
	.....	...	.....	Oktavbereich
	C♯5/D♭5	53	cis''/des''	
	C5	52	c''	
	B4	51	h'	
	A♯4/B♭4	50	ais'/b'	
4(one-lineoctave)	.....	...	.....	Eingestrichener
	.....	...	.....	Oktavbereich
	C♯4/D♭4	41	cis'/des'	
middle C	C4	40	c'	
	B3	39	h	
	A♯3/B♭3	38	ais/b	
3 (small octave)	.....	...	.....	Kleiner
	.....	...	.....	Oktavbereich
	C♯3/D♭3	29	cis/des	
	C3	28	c'''	
	B2	27	H	
	A♯2/B♭2	26	Ais/B	
2 (great octave)	.....	...	.....	Großer
	.....	...	.....	Oktavbereich
	C♯2/D♭2	17	Cis/Des	
	C2	16	C	
	B1	15	,H	
	A♯1/B♭1	14	,Ais/,B	
1 (kontra octave)	.....	...	.....	Kontra-
	.....	...	.....	Oktavbereich
	C♯1/D♭1	5	,Cis/,Des	
	C1	4	,C	
	B0	3	„H	
0 (sub-kontra do-main)	A♯0/B♭0	2	„Ais/„B	Subkontra-Bereich
	A0	1	„A	

The double-sharp elevates a note by two half tones. In German the suffix -isis is attached to its name. The double-flat lowers the pitch by two half tones (the suffix -eses is added to the note in German).

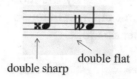

If an accidental is put at the beginning of the first line, it remains valid for the whole duration of the musical piece. Examples are shown below.

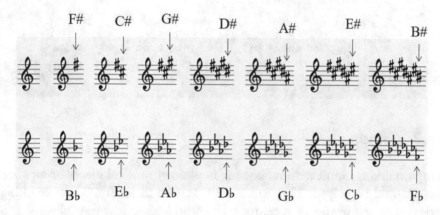

Except B♯ and E♯ (C♭ and F♭) all notes above are played on black keys of the piano.

2.3.7 Loudness

Short letters below the lines suggest the loudness[2] of the passages to be played. They are:
signs for "loud" (with increasing loudness from left to right):
—*mf* (mezzoforte), *f* (forte), *ff* (fortissimo), *fff* (fortefortissimo),
and signs for "soft" (with decreasing loudness from left to right):
—*mp* (mezzopiano), *p* (piano), *pp* (pianissimo), *ppp* (pianopianissimo).
Note that these sings indicate relative differences in loudness: there is no relationship to the well defined value of the loudness measured in dB.

---

[2] Loudness is a term for the sound intensity often used in psychoacoustics.

When the "Hammerklavier" was invented around 1700, it offered the option to change the loudness between piano and forte, by different strength of striking the keys. Due to this dynamic advantage the further developed instruments were called "pianoforte" or later just by the short name "piano".

Symbols indicating dynamic changes in loudness are the crescendo-fork and the decrescendo (or diminuendo)-fork.

crescendo   decrescendo

Coffee Break

Very seldom the marks *ffff* or *pppp* appear. How shall I play in such cases? Should I suffer from auditory disturbance after playing *ffff*? Does anybody hear something from *pppp*? Good conductors know how to manage with *ffff*, as well as *pppp* (see Chap. 10).

2.3.8 Tempo

Tempo markings indicate how fast (or how slow) musical pieces or a part of them should be played. They are usually written in Italian. These markings appeared at first in the 17th century. But who knows how fast "allegro" is or how slow "largo" is? The real tempo depends on the musician's feeling and, therefore, is subjective.

Since the metronom was invented by J. N. Mälzel in 1815, tempo has become more objective.[3] A metronome is a device that produces a sound at a regular interval which can be set by the user, typically in beats per minute (see Fig. 2.8). Nowadays the suggested number of beats per fourth note (or any other note) is often indicated at the beginning of the piece. Figure 2.8 shows a traditional mechanical metronome. As a modern version there exist electronic metronomes with a quartz crystal.

Coffee Break

A very diligent amateur pianist bought a metronome and started exercising with it. Soon after he found something wrong with his new metronome. His metronome accelerated during playing Etudes, while it slowed down during Nocturnes.

---

[3] A prototype of the metronome had been invented by Andalusian polymath Abbas Ibn Firnas in the 9th century.

Fig. 2.8 A traditional mechanical metronome. Tempo is indicated in beats per minute and adjusted with the pendulum (Nikko brand, CC BY-SA 3.0 by Vincent Quach)

In Table 2.3 we show some tempo markings, terms for tempo change, and additional descriptions for tempo, which are often used.

2.3.9 Ornaments

Ornaments modify the pitch pattern of individual notes without changing the overall line of the melody (or harmony). They serve as an acoustic decoration. Typical marks of ornaments and their ways to be played are shown in Fig. 2.9.

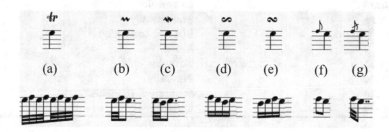

Fig. 2.9 a trill, **b** upper mordent, **c** lower mordent, **d** turn, **e** inverted turn, **f** appoggiatura, **g** acciaccatura. Bottom lines show the way of playing

(a) trill: a rapid alternation between the specified note and the next higher note within its duration

(b) upper mordent: the principal note and then the next higher note are played rapidly followed by the principal note

(c) lower mordent: the principal note and then the next lower note are played rapidly and return to the principal note

Table 2.3 Basic tempo markings, terms for tempo change, and some additional descriptions for tempo

Italian description	Meaning
Basic tempo markings	
Largo	Slowly
Larghetto	Rather broadly (faster than Largo)
Lento	Slowly
Adagio	Calmly,unhurriedly, with great expression
Andante	At a walking pace
Moderato	At a moderate speed
Allegretto	Moderately fast
Allegro	Fast and bright
Vivace	Lively and fast
Presto	Very, very fast
Terms for tempo change	
accelerando	Speeding up
rallentando, ritardando	A gradual slowing down
rubato	Free tempo
a tempo	Back to the original tempo
alla breve	Twice as fast as normal
Additional descriptions used together with tempo markings	
agitato	Agitated
assai	Much
cantabile	In a singing style
con brio	With spirit/vigor, or brightly
dolce	Sweetly
espressivo	Expressive
grazioso	Gracefully
maestoso	Majestically
ma non troppo	But not too much
più	More
poco a poco	Little by little
scherzando	Playfully
sostenuto	Sustained, stretched out
tranquillo	Quietly

(d), (e) turn: a sequence of upper auxiliary note, principal note, lower auxiliary note and principal note or a sequence of lower auxiliary note, principal note, upper auxiliary note and principal note

(f) appoggiatura*: the duration of the first half of the principal note has the pitch of the grace note (written before the principal note) and the second half the pitch of the principal note

(g) acciaccatura**: the grace note is played with very brief duration.

* in German: der Vorhalt, or der lange Vorschlag
** in German: der Vorschlag, or der kurze Vorschlag

2.3.10   Note Relationships

Sometimes arcs or zig-zag lines appear in musical scores. They start from one note and end at the next, or at a further separated note (with several notes in between). These curves (or lines) indicate note relationships, referring to the way in which two or more notes are connected. Typical examples are (see also Fig. 2.10):

(a) tie: two notes of identical pitch played as one note with their durations added together
(b), (c) slur: two or more notes played in one physical stroke
(d) glissando: a continuous and unbroken glide played from one note to the next one for pitches in between
(e) arpeggiated chord: notes of the chord played rapidly (usually ascendingly)

(a)   (b)   (c)   (d)   (e)

Fig. 2.10 **a** tie, **b, c** slurs, **d** glissando, **e** arpeggiated chord

2.3.11   Articulation Marks

Articulations determine specified notes in terms of the length of their sound and the shape of its attack and decay. Some of them have marks as shown in Fig. 2.11, and others are simply described without symbols (an example: "legato").

(a) staccato: note played shorter than nominated, but without shortening the nominated duration
(a') portato: a smooth, pulsing articulation as mezzo-staccato

(b) accent: note played louder with a harder attack
(b') sforzando: note played with sudden, strong emphasis
(c) tenuto: note played at its full value of duration
(d) marcato: note played louder or more forcefully than the surrounding notes
(e) fermata: note (or chord, rest) sustained longer than the nominated duration
(f) legato: notes played smoothly and connected

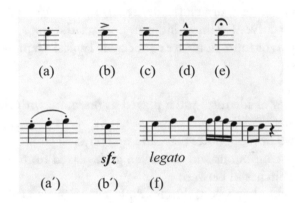

Fig. 2.11 a staccato, **a'** portato, **b** accent, **b'** sforzando, **c** tenuto, **d** marcato, **e** fermata, **f** legato

2.3.12 Other Expressions

Since the 19th century composers have frequently used expression marks in their native language: German, French, English, Russian or others, grasping to some extent the emotional contents and characteristics of the musical piece in question. Below are a few examples:

- *Lustig im Tempo und keck im Ausdruck*, G. Mahler, Symphony No. 3, 5th movement
- *Klärung und Rückschritt*, A. Schönberg, 1. Kammersymphonie
- *Les parfums de la nuit*, C. Debussy, Ibéria-Suite, 2nd movement
- *Boisterous Bourrée, Playful Pizzicato, Sentimental Sarabande, Frolicsome Finale*, B. Britten, Simple Symphony
- *In the silence of the monotonous steppes of Central Asia is heard the unfamiliar sound of a peaceful Russian song* (original in Russian), A. Borodin, Symphonic poem "In the Steppes of Central Asia"

- **Der schwer gefaßte Entschluß**, L. v. Beethoven, String Quartet Op. 135, final movement.

2.3.13   An Example—Rondo of Haydn

The example shown in Fig. 2.12 illustrates some basics of modern musical notation. On the upper left we find the tempo marking "Allegro" with a instruction for the metronome (126 beats per minute for a fourth).

Fig. 2.12 The beginning of "Rondo" from Sonata in D major by Joseph Haydn (Hoboken XV:37)

The line notation starts with the clef: G-clef for the upper (usually for the right hand), F-clef for the lower one (for the left hand). Further to the right the accidentals are written for each of the clefs, meaning that the note F is increased to F♯ and C to C♯. This points to D major as the governing key. The type of rhythm is determined by the numbers 2 over 4, i.e., fourth per bar played with 2 beats. The melody does not start with an emphasized first beat of a bar, but with a less accentuated single note (upbeat).

The **p** between the two lines refers to the dynamics. It means "piano (low intensity)", and is slightly altered by a crescendo and decrescendo-fork.

In the following lines of notes there are bows called "slur", indicating that the notes it embraces are to be played smoothly. Points above certain notes require a staccato, which is a short accentuation.

In the lower line one has double notes and breaks underlining the rhythm of the melody.

At the end one sees the (end) repeat bar, indicating that the part from the beginning to this mark should be repeated.

References

1. R.R. Fay, A.N. Popper (eds.), *Comparative Hearing: Mammals* (Springer, New York, 1994)
2. L.S. Lloyd, A. Fould, International standard musical pitch. J. R. Soc. Arts **98**, 74–89 (1949)
3. F. Gribenski, Plenty of pitches. Nat. Phys. **16**, 232 (2020)
4. H. Helmholtz, *Die Lehre von den Tonempfindungen als Physiologische Grundlage für die Theorie der Musik* (Friedrich Vieweg und Sohn, Braunschweig, 1863)

3

Intervals and Scales

3.1 Intervals

3.1.1 Basic Intervals

Intervals are distances in pitch between two tones, quantified by the ratio of frequencies and labeled by Latin ordinal numbers.

The first eight intervals on the fundamental note C are called (German names in brackets): prime (Prime), second (Sekunde), third (Terz), fourth (Quarte), fifth (Quinte), sixth (Sexte), seventh (Septime), and octave (Oktave). Here, "second" means that the upper tone lies on the second position counted from the line of the first tone (in this case C). If the first tone lies on a line, the second occupies the adjacent space in-between lines. The "third" means the next tone lying on the third position counted from the line of the first tone, and further on. This counting does not reflect the real frequency distance. Therefore, there are some different "second"s and different "third"s and so on.

In the following graphics these tones are built up on the fundamental tone of the C major scale. These can be grouped as perfect and major intervals. In a more detailed way one writes intervals according to the following table (Table 3.1).

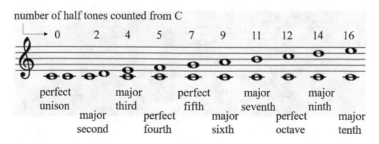

number of half tones counted from C

0 2 4 5 7 9 11 12 14 16

perfect unison — major third — perfect fifth — major seventh — major ninth

major second — perfect fourth — major sixth — perfect octave — major tenth

© Springer Nature Switzerland AG 2021
K. Tsuji et al., *Physics and Music*,
https://doi.org/10.1007/978-3-030-68676-5_3

Table 3.1 Intervals (number of half tones) in English and German

Interval	English	Status	German
0	perfect unison	PC[a]	Prime
1	minor second	D[b]	kleine Sekunde
2	major second	D	große Sekunde
3	minor third	IC[c]	kleine Terz
4	major third / diminished fourth	IC	große Terz
5	perfect fourth	D/IC	Quarte
6	augmented fourth	D	übermäßige Quarte
	diminished fifth	D	verminderte Quinte
7	perfect fifth	PC	Quinte
8	minor sixth / augmented fifth	IC	kleine Sexte
9	major sixth	IC	große Sexte
10	minor seventh	D	kleine Septime
11	major seventh	D	große Septime
12	perfect octave	PC	Oktave
13	minor ninth	D	kleine None
14	major ninth	D	große None
15	minor tenth	IC	kleine Dezime
16	major tenth	IC	große Dezime
...	and further		

[a] perfect consonance, [b] dissonance, [c] imperfect consonance

3.1.2   Other Frequently Used Intervals and Chords

In the building of musical scales, usually within one octave, a certain number of intervals and chords play a specific role, in addition to those introduced above. We often use "Cent" as a unit for intervals, as explained in the next chapter Sect. 4.2.5.

3.1.2.1   Hiatus

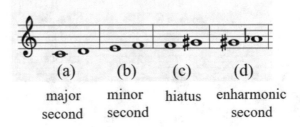

(a) major second

(b) minor second

(c) hiatus

(d) enharmonic second

The second as musical half tone is counted among the dissonant intervals. It can create strong harmonic or melodic tension. The minor second is perhaps the strongest dissonance of all. It tends to sound quite unpleasant and, in certain styles, had to be transformed to consonant intervals. The major second (a whole tone) determines most steps in heptatonic scales (see Sect. 3.2.4). Second steps also form the basic elements for ornaments in baroque and classical music, including the various kinds of trills. The hiatus (augmented second) plays a characteristic role in harmonic minor and Gypsy scales (details see Sect. 5.2.1).

3.1.2.2 Ditonus

The ditonus denotes an interval of two whole tones.

In Pythagorean tuning the ditonus is known as Pythagorean major third. This third is higher than the pure major third (5:4) by a syntonic comma (see Sect. 4.1) and, therefore, sounds a little sharper. In antique music theory the ditonus was generally considered to be a dissonance, but there developed some disagreement, whether one should count it among the consonant intervals.

3.1.2.3 Tritone

The tritone (or tritonus) is an interval encompassing three whole tones. Occasionally, it is called half octave, because in the equal temperament (see Sect. 4.2.4) the tritone cuts the octave into exactly two halves, i.e., 600 of 1200 Cents. Examples are the intervals from C up to F♯ with adjacent whole tones C-D, D-E and E-F♯, or from F up to the B above it (F-G, G-A and A-B). The tritone is considered as an augmented fourth, a chromatic variant of the perfect fourth, and thus does not belong to the diatonic intervals. Or if, for example, the tritone C-F♯ is replaced by the enharmonically equivalent interval C-G♭, it is called a diminished fifth.[1]

Due to technical and harmonical problems, especially in song, the tritone was also called "diabolus in musica" (devil's interval). In fact, since the tritone is a very unstable interval, it was often avoided or composed with immediate ways for dissolving the tension by a lead-note. This proves to be useful for playing intricate modulations.

This interval has been included in many compositions of classical and modern pop music. In Gypsy scales a tritone is the characteristic interval.

---

[1] An enharmonic equivalent is a note, interval, or key signature that is equivalent to some other note, interval, or key signature, but "spelled" or named differently. For example, in any twelve-tone temperament C♯ and D♭ are enharmonic.

It is often used in jazz and blues music, in particular in connection with the blue note (see Sect. 5.7.2). Its reputation as a devil's interval has certainly promoted a presentation of darkness, sorrow, unearthly or demonic moods.

Two examples are shown below. The tritones are marked with red circles.

La dance macabre of Camille Saint-Saëns

Maria in West Side Story of Leonard Bernstein

Richard Wagner used the tritone in his opera "The Flying Dutchman", too. In the whole-tone music (see Sect. 3.2.3) the tritone also appears: in "Baba-Yaga" of Mussorgsky's "Pictures at an Exhibition", for example.

3.1.2.4 Triad

A triad in the classic sense is a harmonic set of three notes that can be stacked vertically in thirds. The triad's members, from lowest-pitched tone to highest, are called **the root**, **the third** and **the fifth** (see below). The interval for the third above the root is a minor or a major third. The interval for the fifth above the third is a minor third or a major third, hence its interval above the root is a diminished, perfect or augmented fifth.

The triad formed on the tonic note of a scale (the first and fundamental note of a scale) is called the tonic. Furthermore, one has the two primary harmonics: dominant and subdominant with G and F, being the roots of the triad as shown below

There are 4 types of triad: (a) the major, (b) the minor, (c) the diminished, and (d) the augmented triad. These are illustrated below for the C as the root. The major triad consists of the root, the major third and the perfect fifth, the minor triad—the root, the minor third and the perfect fifth, the diminished triad—the root, the minor third and the diminished fifth and the augmented triad—the root, the major third and the augmented fifth. The chord can be inverted: (e) and (f) are the first and the second inversion of the major triad, respectively.

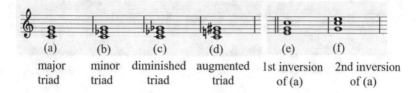

(a)	(b)	(c)	(d)	(e)	(f)
major triad	minor triad	diminished triad	augmented triad	1st inversion of (a)	2nd inversion of (a)

The root tone of a triad, together with the diatonic scale to which it corresponds, sets its basis. The triad's function is determined by its quality: major, minor, diminished or augmented. Major and minor triads are the most commonly used triad qualities in European classical, popular and traditional music.

In a broader sense the term triad refers to any combination of three different pitches, regardless of the intervals amongst them. (Then, the word "trichord" is often used). For example, Schubert uses a triad of the root, the minor third and the perfect fourth (see (a) below) or another triad of the root, the major third and the perfect fourth (b). Mussorgsky uses narrower triads like the root, the minor second and the perfect fourth (c), or the root, the minor second and the minor third (d).

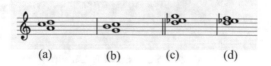

(a)	(b)	(c)	(d)

3.1.2.5 Tetrad

A tetrad is a set of four notes in music theory. When these four notes form a tertian chord (3 thirds stacked together), they are called by standard a seventh chord, because the upper-most note is a seventh with respect to the root of the chord (see the next figure). In the music of the 20th and the 21st century four-note chords are often formed of intervals other than thirds. For instance, a tetrad may contain a sixth, i.e., the interval with the root is a sixth (major or minor). It is then referred to as a sixth chord or as added sixth chord since Jean-Philippe Rameau—"sixte ajoutée" [1]. There are more variants of tetrads, e.g. with an added ninth or a dominant thirteenth. The latter is a chord with root, third, seventh and thirteenth, with the thirteenth designating the interval of an octave plus sixth.

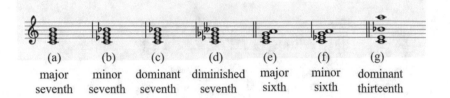

(a)	(b)	(c)	(d)	(e)	(f)	(g)
major seventh	minor seventh	dominant seventh	diminished seventh	major sixth	minor sixth	dominant thirteenth

If not specified otherwise, a "seventh chord" often means a dominant seventh chord: a major triad together with a minor seventh. However, a variety of sevenths may be added to a variety of triads, resulting in many different types of seventh chords. In general, the seventh destabilized the triad, and allowed the composer to emphasize movement in an intended direction. As in time the collective ear of the western world became more accustomed to dissonance, the seventh was allowed to become a part of the chord itself.

Just listen to an excerpt from Chopin's Mazurka in F minor, Op. 68 with dominant sevenths marked in red.

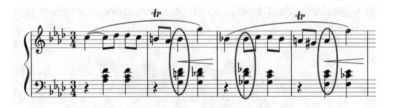

Mazurka Op. 68 of F. Chopin

3.1.2.6 Tetrachord

A tetrachord is a sequence of four tones within a pure fourth [2]. In European music theory one mainly considers the diatonic tetrachord, composed of two whole tones and one half tone with variants: whole-whole-half-tone, whole-half-whole-tone, and half-whole-whole-tone. This scheme offers a possibility to construct the classical diatonic scales by adding a whole note and composing the entire scale by two tetrachords.

In the framework of music didactics this approach is not without problems. For instance, in a minor scale the two contained tetrachords are structured differently. There are church modes (see Sect. 3.2.4) in which the lower tetrachord includes a tritone, which is in opposition to the original Greek ideas of the employment of consonant scales. In Gypsy scales one even has to implement "chromatic" tetrachords consisting of the pattern: half tone - hiatus - half tone.

The joining of tetrachords had been a basic goal of Greek music theory and one finds "recipes" by philosophers and mathematicians like Aristoxenos and Euklid. Besides the diatonic tetrachord they discussed a chromatic one with steps: minor third, half tone, half tone and an enharmonic one: major third, quarter tone, quarter tone. Quarter tones are used, for instance, in Arabic music (see Sect. 5.2.4).

There are written documents on such tetrachords:

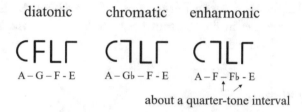

In this description, the instrumental Greek notation uses an ancient Phoenician alphabet. Notes are written in a sequence from left to right, corresponding to the pitch sequence from high to low (other than we do it today).

3.1.2.7 Hexachord

The hexachord is a scale of six successive tones (presumably derived from the six-stringed lyre) developed by Guido da Arezzo.

In the Middle Ages hexachords offered the basis for learning Gregorian songs. Hexacords are an extension of the Greek tetrachord (E-F-G-A), which in the 9th century was transposed downward by one note to the four fundamentals of the church modes (D-E-F-G). Below and above a whole-note step was then added (C and A). In each hexachord a half-tone step separates the middle tones, all others are a whole tone apart.

From the Middle Ages to the Renaissance, music theorists and composers thought about music in six-note hexachords. If the range of a melody was larger than a sixth, they just switched to another hexachord. Hexachords always involved the same diatonic scale structure: Whole-whole-half-whole-whole; asymmetrical hexachords were used in modern music, for instance by Schönberg (see Figure below, left).

Es C H B E G

"Schönberg hexachord" (6-Z44) and "Schönberghexachord" at "EsCHBEG"
transposition and ordering

The right figure is known as Arnold Schönberg's signature hexachord [3]. The transposition contains the pitches Es, C, H, B, E, G (Schönberg), E♭, B, and B♭ being Es, H, and B in German. Schönberg used the hexachords in the song "Seraphita" (Op. 22 No. 1) and the monodrama "Die glückliche Hand (The Hand of Fate)". 6-Z44 is associated with the character of the Hauptmann (captain) in Alban Berg's opera "Wozzeck". It is also one of the fundamental harmonies in the last movement of Igor Stravinsky's The Rite of Spring, "Sacrificial Dance".

3.1.2.8 Comma

Two important intervals in connection with the superposition of perfect fifths, the Pythagorean comma (interval of 23.46 Cent) and the syntonic comma (21.51 Cent) will be introduced in Chap. 4.

The difference between the Pythagorean and the syntonic comma is called **schism** and has the value:

$$\text{schism} = 23.46\,\text{Cent} - 21.51\,\text{Cent} = 1.95\,\text{Cent}.$$

Andreas Werckmeister, in constructing his well-tempered intonation, very carefully considered this small but important difference [4].

3.1.2.9 Limma

Since Euklid this term has been used for the Pythagorean half tone, as it appears in the diatonic Pythagorean tuning (see Chap. 4). It corresponds to the proportion equation:

$$\text{limma} = \frac{\text{fourth}}{\text{ditonus}} = \frac{256}{243} \approx 90.22\,\text{Cent}.$$

3.1.2.10   Apotome

Apotome designates in the Greek antique the Pythagorean chromatic half tone, which is the difference between whole tone and limma, with a frequency relation to be calculated as

$$\frac{(9/8)}{(256/243)} = 113.69 \text{ Cent},$$

thus being by a Pythagorean comma (23.5 Cent) larger than the limma.

3.1.2.11   Diesis

This used to be the name for all intervals smaller than a half tone (introduced by Aristoxenos), but a more specific definition came up later [5].

- Small: If 3 major thirds are put in row, one obtains an octave in equal temperament, but a slightly smaller interval in pure tuning. The difference to the octave is called "small diesis". The frequency difference is determined by the relation

$$\frac{(2/1)}{(5/4)^3} = \frac{128}{125} = 1.024 \approx 41 \text{ Cent}.$$

- Large: Should 4 minor thirds be in a row this results, in equal temperament, in an octave. In pure tuning the interval is somewhat larger, the difference being calculated as

$$\frac{(6/5)^4}{(2/1)} = \frac{648}{625} = 1.0368 \approx 63 \text{Cent}.$$

3.1.3   Beats

A beat is caused by an interference phenomenon, when two sounds of slightly different frequencies (f_1 and f_2) are played simultaneously. The audible frequency is the difference of the frequencies of the original two sounds.

The sound wave 1, $S_1(t)$, and wave 2, $S_2(t)$, are described as:

$$S_1(t) = A_1 sin(2\pi f_1 t + \varphi_1) \tag{3.1}$$

and

$$S_2(t) = A_2 sin(2\pi f_2 t + \varphi_2) \tag{3.2}$$

If the amplitude and the phase of these two sounds are the same, the sound to be heard $S(t)$ is

$$S(t) = Asin(2\pi f_1 t) + Asin(2\pi f_2 t)$$
$$= 2A cos\left(2\pi \frac{f_1 - f_2}{2}t\right) \cdot sin\left(2\pi \frac{f_1 + f_2}{2}t\right), \tag{3.3}$$

where $A = A_1 = A_2$ and $\varphi_1 + \varphi_2 = 0$.

When

$$|f_1 - f_2| \ll (f_1 + f_2),$$

the cosine term $cos\left(2\pi \frac{f_1-f_2}{2}t\right)$ is too low to be perceived as an audible tone. Instead, it is perceived as a periodic variation in the amplitude of the sine term $sin\left(2\pi \frac{f_1+f_2}{2}t\right)$ as illustrated in Fig. 3.1.

The lower frequency cosine terms (the red curves) form an envelope for the higher frequency sine term (the blue curve). The frequency of the envelope is $|f_1 - f_2|$ and this is called the beat frequency.

Musicians use such beats for tuning their instruments.

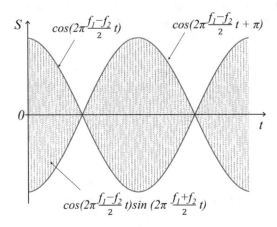

Fig. 3.1 Blue curves: high frequency sine term, red curves: low frequency cosine term in Eq. 3.3

3.2   Scales

3.2.1   About Scales

A scale is a series of tones following well-ordered pitches. In most cases its extension over one octave is considered. Thereby specific rules are followed.

We are already acquainted with an important example: the C major scale, shown here with the G-clef and going up and down again.

It comprises the natural notes (first seven notes, starting with fundamental C). It does not need any accidental. The detailed architecture is determined by the tonal system (diatonic in this case, see Sect. 3.1.2). The upward series covers notes up to the beginning of the next octave.

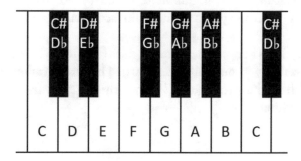

On the piano all white keys within one octave have to be played, all of them are at a distance of a whole note except for the half tones E-F and B-C. The black keys (not belonging to this scale) fill the half-tone gaps between the white keys. Counting all the half-tone steps would lead to the number 12.

Most common scales in and outside Europe are based on 5, 6 or 7 tones within an octave (with some exceptions), the C-major scale having 7 tones/natural notes. We will present in this Section pentatonic, hexatonic, diatonic and chromatic scales. Other more exotic examples from Gypsy scales, Chinese pentatonics, Okinawa pentatonics, Indian ragas, or North-American blues and others are introduced in Chap. 5.

In the following subsections the brackets shown below are used to indicate the type of interval between subsequent tones:

$\bigwedge$: minor second (1 half-tone interval)

no sign: major second (1 whole-tone interval)

$\overline{\llcorner}$: minor third (1 ½ whole-tone interval)

$\overline{}$: major third (2 whole-tone intervals)

3.2.2 Pentatonic Scale

Five tones constitute a pentatonic scale within an octave, thus they have a relatively simple structure. Such scales have developed in many cultural environments over very long time periods. Pentatonics can be found in song, tunes, intonation of musical instruments of many people all over the world. Frequently it is the basis of children's songs, folklore, and also in classical and modern compositions. In our European reference frame there are, in principle, two kinds: the major- and the minor-variant, with the major-variant being applied more frequently.

A pentatonic scale is "preprogrammed" on the piano: just play the five black keys within an octave, and you hear the major scale starting on the fundamental note C♯.

One distinguishes between hemitonic scales (including half-tone steps), and anhemitonic scales (without half-tone steps). The major-variant is an anhemitonic scale with 3 whole tones and 2 minor thirds. Note that the major triad (C-E-G) is included.

C D E G A C

This anhemitonic scale is used for popular compositions of many indigenous people in Asia, Africa, America and early Europe. It is a predecessor of the Greek heptatonics. A world-wide known example is a German children's song: "Laterne, Laterne, Sonne, Mond und Sterne...". For hemitonic scales see Min-you and Miyako scales shown in Fig. 5.11.

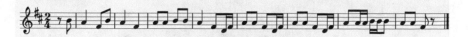

Laterne (German children song)

In the minor variant one detects the minor triad C-E♭-G.

C E♭ F G B♭ C

Two very nice examples of the pentatonic music are cited below:

"Morning Mood" from Peer-Gynt-Suite No.1 Op. 46 of Edvard Grieg

2nd movement of Symphony No. 5 (9) in E minor, Op. 95 of Antonín Dvořák

So many additional and diverse forms and rhythms of pentatonics have been created in different cultural environments (see Chap. 5).

3.2.3 Whole-Tone Scale

Moving from a 5-tone to a 6-tone scale we consider a sequence of whole tones within an octave.

Without any half tone and only intervals of whole tones, the scale consists of 6 steps. There are only two variants of this scale when based on the fundamental C4 or C4♯.

The scale sounds for us relatively unusual, there are no leading notes as in diatonic scales (see Sect. 3.2.4). Therefore, this scale causes a somewhat "floating" impression.

Early use of the whole-tone scale was made by Russian composers: Michail Glinka (1804–1857), Nikolai Rimsky-Korsakov (1844–1908) and Modest Mussorgsky (1839–1881). A well known masterpiece of Rimsky-Korsakov is his symphonic poem "Scheherazade" of the year 1888. It is based on the tales of "Thousand and one Night" and is specific for a colorful instrumentation and the widespread interest for oriental culture. There is the tyrannic Sultan who is comforted every night by a tale told by Princess Scheherazade, who thus escapes punishment day by day.

The initial theme for the sultan uses a descending whole-tone scale: E-D-C-A♯ on the first beat of each measure.

Sultan's theme in Schenerazade of Nikolai Rimsky-Korsakov

Soon afterwards a violin accompanied by a harp plays the leitmotif of the storyteller herself, a tender melody in Dorian church mode (see Sect. 3.2.4) on A.

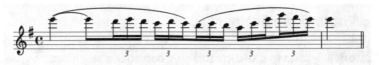

Princess's theme in Scheherazade of Nikolai Rimsky-Korsakov

In one of his most popular compositions, the "Pictures at an Exhibition" for piano, Modest Mussorgsky payed homage to the artist Victor A. Hartmann, who had been a close friend of his. In countless sketches and paintings Hartmann captured the poetry of everyday life and created a gallery of eloquent human figures. Mussorgsky was concerned by the social problems of his time and followed three principles: man—life—truth. Similar views were also shared by the "Mighty Handful" to whom he belonged together with Borodin, Rimsky-Korsakov, Cui and Balakirew.

Mussorgsky managed to transpose the visual effects of Hartmann's paintings into acoustically perceptible pieces. In some parts he used the concept of hexatonics, as in the "Pictures at an Exhibition", picture No. 9 "The hut on chicken's legs (Baba-Yaga)".

The hut on chicken's legs (Baba-Yaga) of Modest Mussorgsky

Mussorgsky had always found inspiration for his works in non-musical impressions. The concept of "absolute music" was foreign to him. He was an autodidactic, highly gifted dilettante, of whom one can right say that his lack of academic musical training had allowed him to create boldly innovative works.

Elements of the whole-tone scale also entered compositions of Claude Debussy (1862–1918), Maurice Ravel (1875–1937) and Olivier Messiaen (1908–

1992), among others. French composers of impressionism were bridging the gap between romantics and modern music.

Debussy began to develop an autonomous harmony connecting European influences with traditional Slavic and Asiatic music, including compositions of Mussorgsky, by using pentatonics and whole-tone scales. The resulting sound patterns were unfamiliar, spherical and reminiscent of paintings by Monet and Gauguin, all of this pointing to the artistic field of impressionism.

In his "En Bateau (In the Boat)" of the "Petite suite" for piano four hands Debussy followed the current French music tradition. This meant: simple harmonics and dancing in elastic rhythm, obeying a music which brings joy.

"En bateau" starts with a part (A) in G major dominated by calm arpeggio cantilenas, followed by a middle part (B) in D major with resolute rhythms. The end of this passage is composed in a whole-tone scale, as shown in the following notes, giving an impression that the boat is still unstable after a storm. All notes are part of the whole-tone scale I, except for the A2 and A3 for the left hand of the secondo (shown by red arrows), which may lead to the G major of the last part (A').

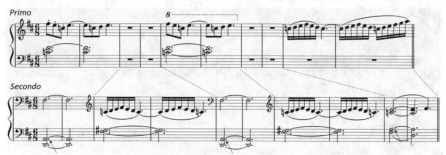

En bateau from Petite Suite of Claude Debussy

Part (A') resumes the arpeggios and leaves us with the gentle motion of the boat on a water surface.

3.2.4 Heptatonics

3.2.4.1 Generalities

If we extend the musical scale by one more note, we form a heptatonic scale consisting of 7 tones. Much of the occidental music is built on heptatonic scales, especially the diatonic one. These were introduced in Greek music theory on the basis of two diatonic tetrachords (see Sect. 3.1.2) and will be discussed in some detail below. Further scales of this kind were used also in non-European music tradition.

A classical scale as a sequence of 7 tones with the fundamental note C is represented by C major, for which all the natural notes are used (see Sect. 3.2.1). The intervals between successive notes consist of 5 whole tones and 2 half tones (E-F and B-C). It is a modern consequence of much earlier efforts to create such diatonic scales, including 5 whole and 2 half tones.

3.2.4.2 Church Modes

Church modes were derived from old Greek music. From the time of Pope Gregor around 600 AD through the Middle Ages (untill the 16th century) they were modified many times to comprise totally 12 (+1) scales: 6 authentic scales (+ Locrian) shown in Fig. 3.2, and 6 plagal scales with the prefix "Hypo". These are lower by 4 tones.

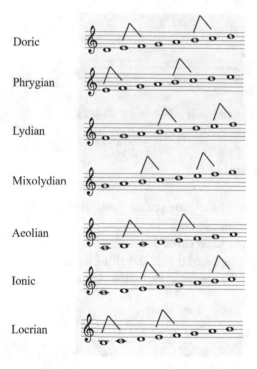

Fig. 3.2 Church modes: six authentic scales and Locrian scale

3.2.4.3 Major Scales

The major scale originates from the Ionic church mode. It consists of 7 tones (heptatonic), intervals of which are 5 whole tones (major second) and 2 half tones (minor second), as shown in the following notes with C as the fundamental.

This can be transposed to other keys. For example, the fundamental of the following scale is D. Two accidentals are required in order to place half tones at the right positions.

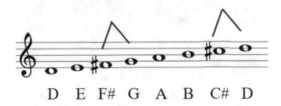

3.2.4.4 Minor Scales

There are three kinds of minor scales: natural minor scale, harmonic minor scale and melodic minor scale.

The natural minor scale came from the Aeolian mode. There are two half tones, one between the second and the third tones, and the other between the fifth and sixth tone. Correspondingly, the A natural minor scale is written as:

natural minor scale on A

Compared to the natural minor scale, the 7th tone of the harmonic minor scale is raised by a half tone, so that the interval between the 6th and 7th tones is one and a half tones (minor third). In the following example, G is raised to G♯.

harmonic minor scale on A

In the upward direction of the melodic minor scale, in addition of raising the 7th tone, the 6th tones is also raised by a half tone to avoid the minor third interval. The downward scale is identical to the natural minor scale.

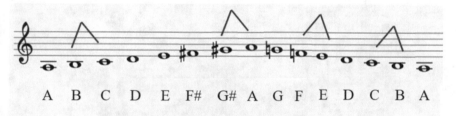

melodic minor scale on A

In Fig. 3.3 we show an example of melodies of a major scale and a minor scale from a lied of Franz Schubert, "Der Lindenbaum (The Linden Tree)". As marked, the minor third and major third are exchanged, and also the places of the minor second are shifted. (For the text of Wilhelm Müller see also Chap. 14).

3.2.4.5 The Circle of Fifth

Figure 3.4 shows the circle of fifth and summarizes the relationship of accidentals and the perfect fifth steps. (Note that this system works only for the equally tempered scale, see Sect 4.2.4). The major scales are written outside the circle with black color, and the minor scales inside the circle with blue color.

In equal temperament the fundamental notes of diametrically opposed keys always form a tritone (the largest harmonic distance).

Fig. 3.3 Lindenbaum of Franz Schubert, top: melody in E major, bottom: melody in G minor

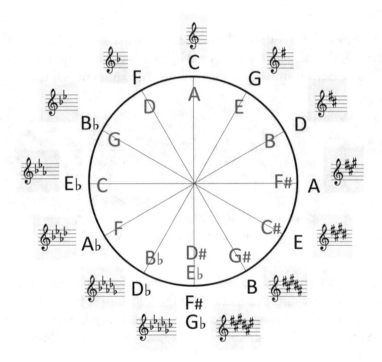

Fig. 3.4 The circle of fifth. Black characters: for major scales, blue: for minor scales

3.2.5 Chromatic Scales

A sequence of all 12 half-tone steps within an octave produces the chromatic scale. If the equal temperament is applied (see Sect. 4.2.4), the structure of subsequent intervals is independent of the tone to begin with. All half-tone intervals are the same. In its notation one has preferred to use, for the upward scale, only ♯ and for the descending part ♭.

Scales like this have been used in classical instrumental music as material for virtuose machinery, thus belonging to the technical basics of all instrumentalists. Its compository usefulness comprises modulations, gliding transitions and glissando effects. The notation of the scale conserves mostly all the diatonic steps and these are not replaced by enharmonically changed tones.

Symmetric Scales

The French composer Oliver Messiaen introduced a series of scales "Modes of limited transposition" [6]. It starts from the whole-tone scale and finishes with the chromatic scale. There are 6 symmetric scales in between.

We show the 5th scale as an example. The tone intervals are: $0.5 - 2 - 0.5 - 0.5 - 2 - 0.5$.

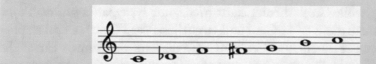

3.2.6 Microtones

Q: Are there musical intervals smaller than a half tone?
A: Of course, there are: from the acoustical point of view any frequency value between two tones defines a tonal step or interval.

Q: Can we hear such a small interval?
A: Yes, but with some restriction, in that our hearing apparatus has a limited frequency resolution, meaning that intervals below that resolution cannot be sorted out (see Chap. 11).

Q: Can I sing melodies in microtones?
A: Yes, because our voice is so flexible and not bound to scales as they have been developed in various cultures for different purposes. We will show in

Chap. 5 such melodies in Indian (shrutis in raga) or Arabian music (macam scales with quarter tones). The term "microtonal music" usually refers to music containing very small intervals around 50 Cent. However, it can include any tuning that differs from European twelve-tone scales. More examples are the Indonesian gamelan music, and music of Thai, Burmese, and African origin. Microtonal variation of intervals is standard practice in the African-American musical forms of spirituals, blues and jazz.

Q: How can we play microtones on musical instruments?
A: Well, that may be often a problem. Keyboard instruments like the piano, instruments with "frets" like the guitar do not provide convenient options for microtones. String instruments like the violin or wind instruments, especially the trombones, open the choice of any frequency for a tone.

As mentioned in Sect. 3.1.2.6, the enharmonic tetrachord in Greek musical theory includes major third, quarter tone and quarter tone. Or more precisely, the enharmonic tetrachord requires a "close-packed" tone sequence—a "pyknon"—where the composite of two smaller intervals is less than the remaining (incomposite) interval. This is a very early introduction of the notion of microtones.

Apparently, while Greek theoreticians tried some systematics on small tone steps (called small diesis ≈ 41 Cent and large diesis ≈63 Cent), a philosopher like Aristoxenos of Tarent (≈ 354–300 BC) declares the human ear as the "highest judge" and calls certain attempts of still finer subdivision of microtones as technocratic and irrelevant.

In 1893 stone fragments, on which Delphic Hymns were inscribed, were found. Figure 3.5 shows a part of one Hymn, which is a vocal notation including a kind of microtones—pyknons.

Although microtones have been very influential in Indian and Arabic culture, they were not used for a long time in Europe, until Alexander John Ellis proposed in the 1880s an elaborate set of exotic just intonation tunings and non-harmonic tunings. Ellis also studied the tunings of non-European cultures and stated that they used neither equal divisions of the octave nor just intonation intervals.

Ellis is the person who produced an English translation of Helmholtz's seminal book "*On the Sensations of Tones as a Physiological Basis for the Theory of Music*" [7]. This book includes an analysis of the harmonic series of tones. The higher the order of a harmonic, the smaller the interval to its neighbors (see Sect. 4.3). So there occur necessarily intervals in the microtone range contributing to microtonal music.

Fig. 3.5 Original stone at Delphi containing a hymn to Apollo. Music notation: line of occasional symbols above the main uninterrupted line of Greek lettering (public domain)

Claude Debussy, when listening to a Balinese gamelan performance at the world exposition in Paris (see about gamelan in Sect. 5.5.1), was impressed by the non-European melodies and rhythms. He subsequently made innovative use of whole-tone scales, quite different from the equal temperament. This may also have influenced his use of bell and gong-like sonorities in some of his piano works. In addition, his interest grew in the microtonal intervals based on Helmholtz's work, as documented in pieces like the "Prélude à l'après-midi d'un faune (Prelude to the Afternoon of a Faun)" or "La mer (The Sea)".

Around 1920, a branch of "New Music" came into being, dealing, among other aspects, with microtonality in electronic music. Electronic music facilitates the use of any kind of microtonal tuning. In 1954, Karlheinz Stockhausen built his electronic "Studie II (Study II)" on an 81-step scale starting from 100 Hz with the interval of 51/25 between steps, and in "Gesang der Jünglinge (Song of the Youths)" (1955–56) he used various scales, ranging from seven up to sixty equal divisions of the octave. Some highly motivated modern composers are now exploiting the possibilities of microtonal music. We just mention Charles Ives, György Ligeti, Ferruccio Busoni. Also Béla Bartók used sporadically microtones. For more details see [8].

Specific symbols for microtones by the use of additional accidentals were introduced in European culture, as shown below.

Left: quarter tone sharp, sharp, three quarter tones sharp; right: quarter tone flat, flat, (two variants of) three quarter tones flat (public domain from Wikipedia:en)

As far as our enjoyment of bright or sad music is concerned, the microtonal approach may leave some expectations behind. So far, it is a great field for music theory, but less so for what our musical mind may expect.

References

1. J.-P. Rameau, *Nouveau Système de musique théorique*, Paris 1726, reprinted Broude Bros, New York, 1965)
2. R. Amon, *Lexikon der Harmonielehre*, 2nd edn. (Doblinger, Wien, 2015), p. 305
3. M.L. Friedmann, *Ear Training for Twentieth-Century Music* (Yale University Press, New Haven, 1990), p. 109
4. D. Bartel, Andreas Werckmeister's final tuning: the path to equal temperament. Early Music **43**, 503–512 (2015)
5. D. Benson, *Music: A Mathematical Offering* (Cambridge University Press, Cambridge, 2006)
6. O. Messiaen, *Technique de mon Langage Musical* (Alphonse Leduc, Paris, 1944)
7. H. Helmholtz, *Die Lehre von den Tonempfindungen als Physiologische Grundlage für die Theorie der Musik* (Friedrich Vieweg und Sohn, Braunschweig, 1863)
8. M. Stahnke, About Backyards and Limbos: Microtonality Revisited, in *Concepts, Experiments and Fieldwork: Studies in Systematic Musicology and Ethnomusicology*, ed. by R. Bader, C. Neuhaus, U. Morgenstern (Peter Lang, Frankfurt a. M. and New York, 2010), pp. 297–314

4

Tunings—From Pythagoras to Equal Temperament

4.1 Pythagorean Tuning

Ancient Greek culture, as shown for the development of well-structured musical scales, has been of utmost importance for our music in the European frame. Among the eminent Greek poets, scientists and artists there was this outstanding philosopher and mathematician Pythagoras (570–510 BC) with many followers, whom school children all over the world know because of his theorem on the triangle. He lived on the island of Samos, then emigrated to Southern Italy.

Pythagoras believed that the entire cosmos is ordered according to harmonic numerical relations and that music is a reflection of this order. So he searched for symbolic connections between numbers and tones. In his musical studies he investigated the length of a cord over a monochord to determine intervals with frequency ratios of 2:1 (octave), 3:2 (fifth), 4:3 (fourth) and 9:8 (whole note), all of which are simple numerical ratios [1]. (Note that he did not determine the ratio for the third yet.)

On this base Pythagoras (or some of his followers) realized that for performing music together an exact choice of (consonant or dissonant) frequency intervals was necessary. From his experiments a circular representation can be created to show all the 12 consecutive pure fifth (frequency ratio 3:2) in one graph (illustrated in Fig. 4.1), similar to the circle of fifth of Fig. 3.4. Starting from C0, the note which is the higher fifth G0 (operation Q1), from G the next operation Q2 (further higher fifth) provides the tone D1 (this D is higher than C1). Cn is the C which is n octaves higher than the initial C0. The operation Q3 leads to A1, ... and the operation Q12 results in B♯6.

© Springer Nature Switzerland AG 2021
K. Tsuji et al., *Physics and Music*,
https://doi.org/10.1007/978-3-030-68676-5_4

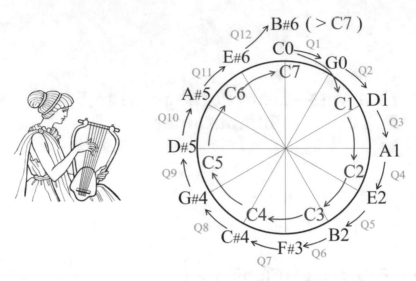

Fig. 4.1 Left: playing the lyre (public domain by Pearson Scott Foresman), right: circle of Pythagorean tuning

This way the Pythagorean tuning was well defined. However, there existed a slight disproportion. In principle, one would expect that the tones B♯6 and C7 are the same, but a difference was found, called the **Pythagorean comma**: How can this be understood?

Just calculate the difference in frequency ratios between 12 fifths and 7 octaves:

$$\frac{(3/2)^{12}}{2^7} = \frac{3^{12}}{2^{19}} \approx 1.01364 = 23.46 \text{ Cent} \approx \text{ one eighth tone.}$$

(The "Cent" is explained in Sect. 4.2.5 below.)

The exact value of this comma was provided later by Euklid around 300 BC [2]. Two great minds were collaborating on a big problem! In practical usage one often chose 11 fifths plus a "wolf fifth", which is slightly smaller than the other pure fifths, in order to close the cycle.

Another frequency-related quantity is the **syntonic comma** [3].

Superposition of 4 perfect fifths produces a tone having a distance from the second octave above the fundamental by a Pythagorean third (which actually corresponds to the **ditonus** described in Chap. 3):

$$\text{Pythagorean third}: \frac{(3/2)^4}{2^2} = \frac{3^4}{2^6} = \frac{81}{64} \approx 407.82 \text{ Cent.}$$

This third is about a fifth of a half tone above the pure third,

$$\text{pure third}: \frac{5}{4} \approx 386.31 \text{ Cent.}$$

(For "pure third" see under "just intonation" in Sect. 4.2.1).
Then, the syntonic comma is

$$\frac{81}{64} \cdot \frac{4}{5} = \frac{81}{80} \approx 21.51 \text{ Cent.}$$

From these considerations the Pythagorean tuning was developed [4]. A scale is based on lengths on the monochord with whole-numbered frequency proportions such as:

- octave (proportion 2:1, half of the length),
- fifth (proportion 3:2, at 3/5 of the length),
- fourth (proportion 4:3),
  because

$$\text{fourth} = \text{octave up } (2/1) \text{ and fifth down } (2/3) \rightarrow \frac{2}{1} \cdot \frac{2}{3} = \frac{4}{3},$$

- whole tone (proportion 9:8),
  because

$$\text{whole tone} = \text{fifth up } (3/2) \text{ and fourth down } (3/4) \rightarrow \frac{3}{2} \cdot \frac{3}{4} = \frac{9}{8}.$$

This procedure does not include the pure third (5:4), but the ditonus (81:64), which is slightly larger by the syntonic comma (81:80). Still missing is the limma as Pythagorean half tone with a ratio of 256/243 ($\approx$ 90.22 Cent, see limma in Sect. 3.1.1). The inclusion of two limma-steps between E-F and H-C concludes the diatonic character of this scale. As easily noticed in the Table 4.1 below, similar two half tones appear in all discussed tunings and have remained a significant characteristics of the C major and any other major (or minor) scale up to our modern music.

On the basis of the fundamental C one would tune the following tones by their quintic distances:

... F C G D A E B...

In diatonic order one obtains the Pythagorean scale as shown in Table 4.1.

This way the fifths and fourths are pure, while the thirds do not exactly correspond to the pure interval (5:4), but differ in the Pythagorean tuning by the syntonic comma.

Table 4.1 Diatonic order of C major in various scales

	C	D	E	F	G	A	B	C
	perfect unison	major 2nd	major 3rd	perfect 4th	perfect 5th	major 6th	major 7th	perfect octave
Pythagorean tuning								
[a]	1:1	9:8	81:64	4:3	3:2	27:16	243:128	2:1
	=1	1.125	1.266	1.333	1.5	1.688	1.90	2
[b]	0	203.9	407.8	498.0	701.9	905.8	1109.7	1200
[c]	WT		WT	L	WT	WT	WT	L
[d]	9:8		9:8	256:243	9:8	9:8	9:8	256:243
[e]	203.9		203.9	90.2	203.9	203.9	203.9	90.2
Just intonation								
[a]	1:1	9:8	5:4	4:3	3:2	5:3	15:8	2:1
	=1	1.125	1.25	1.333	1.5	1.667	1.875	2
[c]	0	203.9	386.3	498.0	701.96	884.4	1088.2	1200
[d]		9:8	10:9	16:15	9:8	10:9	9:8	16:15
[e]		203.9	182.4	111.7	203.9	182.4	203.9	111.7
Meantone temperament								
[b]	0	193.2	386.3	503.4	696.6	889.7	1088.9	1200
[e]		193.2	193.2	117.1	193.2	193.2	193.2	117.1
Equal temperament								
[a]	$\sqrt[12]{2^0}$	$\sqrt[12]{2^1}$	$\sqrt[12]{2^4}$	$\sqrt[12]{2^5}$	$\sqrt[12]{2^7}$	$\sqrt[12]{2^9}$	$\sqrt[12]{2^{11}}$	$\sqrt[12]{2^{12}}$
	=1	≈1.1224	≈1.2599	≈1.3348	≈1.4983	≈1.6818	≈1.8877	=2
[b]	0	200	400	500	700	900	1100	1200
[e]		200	200	100	200	200	200	100

[a] frequency relation with fundamental C, [b] frequency interval from fundamental C in Cent,
[c] WT=whole tone, L=limma (Pythagorean half tone), [d] frequency relation with neighboring tones,
[e] frequency difference from neighboring tones in Cent

Coffee Break

A legend has been passed down over the centuries saying that Pythagoras once listened to a blacksmith and noticed harmonies in the tones of the four hammers there. The consonances or dissonances would depend on the weight of the hammers, when they were struck simultaneously. The hammers A, B, C and D weighed 12, 9, 8, and 6 pounds, respectively. Hammers A and D were in a ratio 2:1, which is the ratio of an octave. The hammers A to B and A to C have ratios 12:9 = 4:3 = perfect fourth and 12:8 = 3:2 = perfect fifth, respectively. The ratio of hammers B and C was 9:8, which is equal to a whole tone.

The legend was later found to be false, at least with respect to the hammers. The hammers, in fact, do not produce such harmonic intervals.

These proportions are, indeed, relevant to the string length on the monochord. With his and his followers' studies on the monochord Pythagoras put forward the historically first empirically secure mathematical description of a physical fact.

4.2   Some More Intonation Systems

4.2.1   Just Intonation (or Pure Tuning)

Pythagorean tuning remained quite common during many centuries. It was adapted by the Romans and, for some time, in the European Middle Ages, practiced also in Gregorian chorals, which profited from its pure intervals.

But changes were due, especially in music where several voices or instruments played together, and when keyboard instruments with less variability for tuning were included.

The first mention of the major third was made in 1300 by W. Odington (*De Speculatione Musices*) [5]. An important achievement was that in addition to pure octaves (frequency ratio 2:1) and fifth (3:2), the major third (5:4) was included. This role was noticed in a wider circle of musicians in 1529 by the work of Lodovico Fogliano [6]. Thus, the just intonation is based on octave, fifth and major third, and all other intervals are composed from these three basic intervals: this is frequently called the "Quint-Terz-System".

The major third had a tendency to sound "rough" (due to the occurrence of a beat), now being considered as a consonant interval. This idea became commonly used with the advent of polyphonic a-chapella-singing or ensemble music of wind and string instruments. The pure tuning creates the feeling of a stable fundamental basis and a reduction of beating. However, intonation by the increased possibilities of chord options make musical life more complicated (but nicer). Fine tuning will not be discussed here, but we add to the Table 4.1 the typical frequency relationships of the just intonation.

A particular feature of this intonation is the existence of two frequency relations for whole tones (major seconds): 9:8 (C-D, F-G) and 10:9 (D-E and G-A). The product of the two whole tones (for example, C-D-E, F-G-A,

G-A-B) is

$$\frac{9}{8} \cdot \frac{10}{9} = \frac{5}{4},$$

corresponding to the major third. This way the pure scales differ clearly from the equal temperament (see below), where each half tone corresponds exactly to a twelfth of the octave.

4.2.2 Meantone Temperament

Other tuning systems came up according to special requirements. The meantone temperament was used in Renaissance and Baroque or even later (until the 19th century), mainly for keyboard instruments with many pure thirds, however only for a restricted number of tonalities. For this purpose the Pythagorean third was transformed to the pure third by reducing the pure fifths by 1/4 of the syntonic comma. In the pure tuning the pure third is embedded into a major whole tone (10:9) and a minor one (9:8), while in the meantone tuning it is divided into two equal whole tones with frequency ratio

$$\sqrt{\frac{9}{8} \cdot \frac{10}{9}} = \sqrt{\frac{5}{4}}.$$

Some frequency relations for a C major scale are added to the Table 4.1, but only in Cents, because the calculation of frequency relations is complicated by the fact, that dealing with the major third leads to irrational frequency numbers.

4.2.3 Well-Tempered Mood

The meantone temperament being a compromise for certain central scales, people in the course of the 17th century wanted to get more flexible in their choices and started to develop systems which could be played in all scales, if not with equal quality. For instance, the purity of the thirds was not always as expected. But one tried to construct scales which had minimum deviations from the just intonation.

These are the well-tempered moods, first suggested by Andreas Werckmeister in 1691 [7]. There was, however, no "temperature" that achieved general acceptance, instead different "temperatures" became popular. A side effect of these efforts was that each of these scales obtained its own sound character (Fig. 4.2).

A major goal, after all, was the creation of a scale in equal steps. Some systems aimed at a relatively clear sound of scales with only few alterations, but still letting music with many alterations sound playable. Other systems tried to make all keys well playable—a predecessor of the later equal temperament—, with the disadvantage of taking a subdued sound into account.

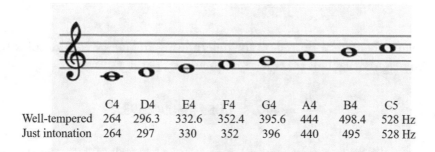

	C4	D4	E4	F4	G4	A4	B4	C5
Well-tempered	264	296.3	332.6	352.4	395.6	444	498.4	528 Hz
Just intonation	264	297	330	352	396	440	495	528 Hz

Fig. 4.2 Comparison between tone frequencies of well-tempered and just intonation

4.2.4 Equal Temperament

Among the well-tempered intonations the equal temperament is the most "radical" one. An early calculation was done by Ch. Tsai-yü, China, in 1584 with a system of 9-digit numbers [8], noticed in Europe only around the year 1800. Another approach stems from Vincenzo Galilei (lute artist and father of Galileo Galilei) in the 19th century, who suggested for half-tone intervals the ratio 17:18 [9]. He shortened strings on his lute by 17/18 of their length. After the 12th half tone he arrived (almost) at the octave. Anyway, the approach is correct in that the octave is subdivided into 12 identical half-tone steps. This way all keys sound the same but to a certain extent not pure any more.

Table 4.2 Comparison between frequencies of just intonation and equal temperament

Chromatic scale of just intonation in C major and C minor completed with F♯ and D♭												
Tone	C	D♭	D	E♭	E	F	F♯	G	A♭	A	B♭	B	C
Freq. Hz	264	281.6	297	316.8	330	352	371.25	396	422.4	**440**	475.2	495	528
In Cent	0	112	204	316	386	498	590	702	814	884	1018	1088	1200
Chromatic scale of equal temperament												
Tone	C	C♯/D♭	D	D♯/E♭	E	F	F♯/G♭	G	G♯/A♭	A	A♯/B♭	B	C
Freq. Hz	261.6	277.2	293.7	311.1	329.6	349.2	370	392	415.3	**440**	466.2	493.9	523.3
In Cent	0	100	200	300	400	500	600	700	800	900	1000	1100	1200

The mathematical procedure for determining frequency relationships between neighboring half tones with frequencies f_1 and f_2 is:

$$\frac{f_1}{f_2} = \sqrt[12]{2} \approx 1.0594630943593.$$

This way, the Pythagorean comma is equaled out. Now all fifth are tuned lower by 1/12 of this comma. The circle of fifths closes completely (see Table 4.1)

Examples:

> perfect fifth: 702 Cent in just intonation $\longleftrightarrow$ 700 Cent in equal temperament
> major third: 386 Cent in just intonation $\longleftrightarrow$ 400 Cent in equal temperament

There is no "ideal", purely tuned interval any more, as shown in Table 4.2.

4.2.5   The Cent

The above comparisons were done in the unit "Cent". We now go back to the origin of this important unit: in 1875 it was proposed by Alexander J. Ellis in an appendix of his translation of Helmholtz "*On the sensations of tone as a physiological basis for the theory of music*" [10]. The Cent is a logarithmic unit of measure used for musical intervals.

Twelve-tone equal temperament divides the octave into 12 half tones of 100 Cents each. Typically, Cents are used to express small intervals, or to compare the sizes of comparable intervals in different tuning systems.

Our auditory system interprets the exponential frequency course as linear, that is, it perceives the logarithm of an actual frequency. In fact, the interval of one Cent between successive notes is too small to be perceived. The smallest resolution of frequency difference for humans is 3–6 Cent (for tones above $\approx$ 1000 Hz) (see psychoacoustics in Chap. 11).

As a standard for the Cent we have

$$\sqrt[1200]{2} \approx 1.0005777895.$$

The Cent system takes care of the fact that with the underlying equal temperature possibly occurring impurities are distributed equally on all tonal steps, i.e. all tonal keys can be used with equal quality. However, this way the individual character of tonal scales may get lost, as often suspected by some composers.

Equal temperament is especially important for instruments like the piano and the organ (no cumbersome retuning). Strings or wind instruments can still tune according to pure principles, since they can adapt differences in tonal height and precision (e.g. for sonatas for violin and piano).

Note that Bach's "Well-tempered Clavier" followed a well-tempered tuning, but not yet the equal temperament in which this work is played nowadays.

Calculations for the Cent

For determining tone frequencies of the whole scale of equal temperament we have the mathematical prescription:

$$f(i) = f_0 \cdot 2^{i/12}, \tag{4.1}$$

where f_0 is an arbitrary starting note (e.g. the chamber tone A4 = 440 Hz) and i is the half-tone distance. This results in a geometrical sequence.

Examples:
G4 is 2 half-tone distance lower from A4, therefore, $i = -2$.
Using Eq. 4.1, we obtain

$$f(-2) = 440 \text{ Hz} \cdot 2^{-2/12} \approx 391.995 \text{ Hz}$$

G5 has 10 half-tone distance higher from A4, meaning, $i = 10$.
Thus we obtain

$$f(10) = 440 \text{ Hz} \cdot 2^{10/12} \approx 783.99 \text{ Hz}$$

G5 has the double frequency of G4. Subsequently, the clean octave distance is conserved.

If intervals follow each other, one can add them, whereas their frequency ratios have to be multiplied.

Examples:

$$\text{fifth} + \text{fourth} = 702 \text{ Cent} + 498 \text{ Cent} = 1200 \text{ Cent (octave)}$$

$$\left(\text{ratios} : \frac{3}{2} \cdot \frac{4}{3} = \frac{2}{1} \right)$$

$$\text{minor third} + \text{ major third} = 316 \text{ Cent} + 386 \text{ Cent} = 702 \text{ Cent (fifth)}$$

$$\left(\text{ratios} : \frac{6}{5} \cdot \frac{5}{4} = \frac{3}{2} \right)$$

Regardless of the particular scale used, one may encounter a beat phenomenon which influences the quality of individual tones. Beating, as described in Sect. 3.1.3, can be heard between notes that are near to a harmonic interval, e.g. between some harmonic of the first note with a harmonic of the second one. For example, in the case of a perfect fifth (e.g. 200 Hz) the third harmonic of the lower note (600 Hz) beats with the second harmonic of the other note (500 Hz), which is (at least) close to 600 Hz. This can also happen for equal temperament intervals, because of differences between them and the corresponding just intonation intervals.

Coffee Break

Pythagoras knew the problem of tuning of "the operation fifth", and he also knew that this problem cannot be solved. A similar situation I can show you in the photographs below. You have to fill a box with various units which have different shape and size (this is a good problem for Penrose [11]). You have to fill the box with a maximum number of units, but nevertheless it should look nice. In the upper box you have so many units, but there is no harmony. The lower box looks better, more harmonic, but there remain a lot of units outside of the box. If some elements are shifted to the right, one or two more units can be added, but it does not look so harmonic. And if some elements are shifted to the bottom,.... Since the Middle Ages, people have been playing this game.

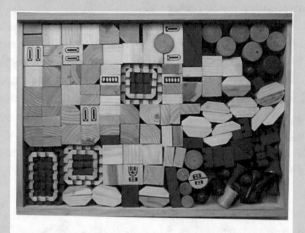

Where is the perfect major third (photos by KT)?

4.3 Harmonic Series

4.3.1 Properties

Through many centuries different musical scales were invented and developed that correspond to our need to identify consonances and dissonances in singing tunes or playing melodies, alone or in groups. We find that many problems to be solved for creating "pure sound" are discussed in the framework of musical perception and mathematics.

Is there some physics behind? And if so, how does it "interfere" with the systematic procedures already presented?

An elastic body (e.g., a thin string or the tube of a wind instrument) may vibrate, upon proper initiation, smoothly on its whole length, but at the same time it may perform certain vibrations at half of its length or less than that (a third, a fourth, and even less,...). For a string transversal extensions occur this way, and for a wind tube, the density of the air is modulated in a similar fashion, both processes coupling to the surrounding air or resonant cavities and thus emitting tones which can be perceived.

The generation of such harmonics of a fundamental frequency of a string is shown in Fig. 4.3.

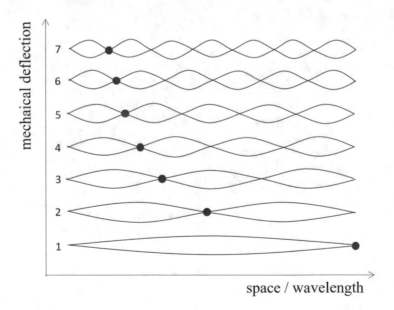

Fig. 4.3 Harmonic partial oscillations of an idealized string

The outcome is the emanation of a whole series of vibrations which create sound at a ground frequency and additional higher frequencies. These are called the "harmonics". For any fundamental tone, let us say the note C, a simultaneous set of harmonics is induced to sound together with the fundamental. Such a set is called the "harmonic series". (In principle, any other note could be chosen as the fundamental.)

Let us consider an example with C2 (65.4 Hz) as the fundamental tone on a string—the first harmonic. Dividing the string into two equal parts leads to the octave with double frequency—the second harmonic. With division into 3 parts we find the third harmonic, which corresponds to the octave of C2 plus a fifth, further on with 4 parts the fourth harmonic with the double octave of C2, called C4.

A whole series of tones (harmonics) will follow this way, as shown in Fig. 4.4. Table 4.3 summarizes the harmonic series of the C2 as fundamental tone up to the 16th harmonic.

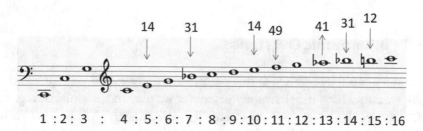

Fig. 4.4 Pitch of harmonics on fundamental tone C2 up to the 16th harmonic. The numbers with upward and downward arrows indicate differences between frequencies in the harmonic series and those of equal temperament in Cent

From this Table we learn that the harmonics of chordophones (like the violin) have frequencies which are (almost) whole-numbered multiples of the fundamental.

Natural wind instruments, which have neither finger holes/valves nor a sliding system to change the length, produce eigenfrequencies directly by the technique of blowing. Such a set of eigenfrequencies is called the "natural harmonic series". These stem from standing waves forming in the instrument according to the total length of the blown tubes. The natural horn (see Sect. 7.5.2), the natural trumpet, and the alpine horn are good examples. Note that the sound we hear belongs explicitly to the overtone series of the natural wind instrument, where only one note is played.

Table 4.3 Some properties of the harmonic series of Fig. 4.4

Harmonic no.	Tone	Frequency ratio	Frequency in Hz
1	C2	1:1	66
2	C3	2:1	132
3	G3	3:2	198
4	C4	4:3	264
5	E4	5:4	330
6	G4	6:5	396
7	Bb4	7:6	462
8	C5	8:7	528
9	D5	9:8	594
10	E5	10:9	660
11	F5	11:10	726
12	G5	12:11	792
13	Ab5	13:12	858
14	Bb5	14:13	924
15	B5	15:14	990
16	C6	16:15	1056

4.3.2 Inharmonic Overtones

If our story could be finished at this point, everybody would be happy and satisfied, we could have nice harmony whenever we play music alone or together.

In practice, however, we hear often slight dissonances. We hear overtones, some of which agree with the harmonics, while others deviate from the harmonics calculated from the frequency ratio. (Exactly speaking, a harmonic is any member of the ideal set of frequencies that are positive integer multiples of a fundamental frequency. On the other hand, an overtone is any partial above the fundamental, regardless of harmonicity. Often the term harmonic is used for the overtone.)

Why do we hear inharmonic overtones and not pure harmonics? There are three factors which we have to take into account. These are the basic problems of tuning systems, instrumental properties and the time window.

4.3.2.1 Problems of Tuning a System

Following Pythagoras no tuning system can realize all harmonics precisely. The numbers with upward and downward arrows written above several notes of the harmonic series in Fig. 4.4 indicate frequency differences between harmonic and equally tempered tone frequency in Cent (leaving away values below 5 Cent which are not well noticed by the human ear). Thus, quite a number of disagreements occur and there is not always a one-to-one correspondence between pitches of the harmonics and a common tonal scale.

Prominent examples:

- the "natural seventh" (7th harmonic 462 Hz) deviates by 31 Cent from B♭4 = 466.2 Hz of the equal temperament (or 49 Cent from B♭4 = 475.2 Hz of the just intonation), for the 14th harmonics there is a comparable situation;
- the "Alpine horn Fa" (11th harmonic = 726 Hz) deviates by 49 Cent from the equal temperament (or 53 Cent from the just intonation), this "Fa" is a quarter tone between the fourth and the tritone;
- for the 13th harmonic (858 Hz) one has a deviation of 41 Cent from A♭5 of the equal temperament.

As seen in Fig. 4.2, just intonation is probably a good compromise, at least the lower harmonics agree with it. Agreement in the lower harmonics is a very important fact, because higher harmonics or overtones are not audible for our ear—their amplitude is small and the frequency itself is out of the range of our audio capacity.

4.3.2.2 Instrumental Properties

If ideal, one can play tones, the frequencies of which are identical to the values listed in Table 4.3, with a violin (or other violin family members), but no piano can manage this. A piano should be tuned according to one of the scales, because each string has a fixed length. Therefore, to play together with a piano causes some inharmonicity. It is of no influence whether a piano is tuned according to equal temperament or just intonation.

Moreover, no instrument can behave like an ideal string (or an ideal tube). Therefore, the frequency of each harmonic has a margin of error probably around ± a few %.

4.3.2.3 Limited Time Duration

Finally, any vibration has limited time duration, which results in a broadening of spectral lines. Figure 4.5 represents an example of the sine wave ($\sin \omega_0 t$). If the sine wave lasts for infinitive time, its spectrum is a single line. On the other hand, if the time is limited, its spectrum is not any more a single line, but has tails both for values smaller than ω_0 and larger than ω_0.

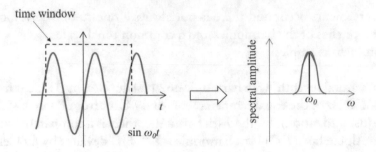

Fig. 4.5 Spectral amplitude of sine curve with time window

4.3.2.4   Additional Comments on Harmonics

Natural tones of wind instruments have some harmonics which are initiated selectively by strong blow (to be compared with the flageolett on string instruments).

Pipes, rods, plates and bells also generate higher frequencies, which are different from whole-numbered multiples (details see Sects. 8.2.1 and 8.2.2). Therefore, sound can be unclear, and sometimes it is recognized as "wrong" (inharmonic).

There are frequency ranges of some width typical for certain instruments and for the human voice, called "formants". These are amplified by specific resonances (as discussed in Sect. 8.3.3).

Even in the case of aerophones and chordophones inharmonicity is caused for the following reasons:

- High string tension for string instruments, very thick strings for double bass, or very short strings
- Existence of noise factors due to strong hits, air blasts, or consonants in singing.

4.3.3   About Tonality

As we have mentioned above, there are discrepancies between pitches of the harmonics and those we hear as an overtone series. Since old Greek time harmonics and overtones have been a big problem for music theoreticians. But why? From the circle of the Pythagorean tuning (Fig. 4.1) it is clear that there is no perfect tuning.

But still, we need some good compromises, if we like to hear "music". The harmonics are not explicitly played notes of a scale or a melody, but "partial" tones radiating together with a given fundamental note, without being perceived individually. The harmonics distribution leads to the timbre of the instrument played—for the sound: brilliant, sharp, radiating, dumb, etc., depending on the number, intensity and pitch of the harmonics, as described in the next section.

In some cultures (e.g., in Mongolia) a singing technique using overtones is common. Singers manipulate the resonances created in the vocal tract in order to produce additional overtones above the fundamental note being sung.

In principle, all intervals of a given scale can be derived from the harmonic series, but some "repair" work (modification with compromises) is necessary.

4.4   Timbre (Color of Sound)

With all these musical and technical considerations about different scales and the harmonic series we will not forget, for a proper ending of this chapter, the most important aspect for which this series of different tones and their harmonics is responsible: the color of sound or its timbre. This comes along as a kind of summary of all the above mentioned effects providing a specific color to each sound of an instrument being played or a voice singing.

In Fig. 4.6 we show wave forms emitted by several instruments: sound waves of guitar, oboe, clarinet, saxophone and trumpet in comparison to the sine wave of a tuning fork. The wave forms differ from each other, although the pitch ($f = 1/T$) is identical. Wave forms of flute, violin and piano are illustrated in Figs. 6.16 and 11.5.

All instruments including multi-dimensional vibrations, like drums and bells, display such characteristic wave forms that determine their sound color or their timbre. Exceptions are, as we know from the previous text, some harmonics played in a singular fashion (natural tones with horns, flageolett with strings).

It has been suggested that human ears and brain tend to group phase-coherent, harmonically-related frequency components into a single sensation. Humans perceive sound waves together as a timbre, and the overall pitch is heard as the fundamental of the harmonic series being experienced [12]. For our reception of music, especially in ensembles, the effects of timbre are vital to appreciate the variability of sound in a melody, a composed music or a tune.

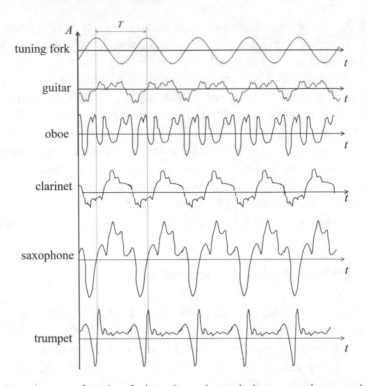

Fig. 4.6 Sound wave of tuning fork, guitar, oboe, clarinet, saxophone and trumpet (from top); ordinate: amplitude in arbitrary units. All waves have the same period T $(= 1/f)$

Intermezzo

Tone Painting

Somebody who likes to learn more about the characteristic sound of certain orchestral instruments would be well advised to listen to the symphonic fairy tale "Peter and the Wolf" (in Russian: Pétya i volk) composed by Sergei Prokofiev in 1936. A narrator tells a story, while the orchestra illustrates it. This work belongs to the most frequently performed ones in the entire classical repertoire.

In the story we make acquaintance with a number of characters, each being musically represented by a specific instrument. There are:

– Peter (the hero) with his violin, walking through the scene with bright and radiating sound,
– the small bird playing beautiful tunes with his flute,
– Grandpa carrying a bassoon and making deep sounds,
– the duck, squeaking according to the tones from an oboe,
– the cat with a clarinet, dangerously observing the scene,
– finally: the giant gray wolf: blowing on a horn, but then caught by Peter.
– Luckily three hunters appear, who use the timpani to shoot the wolf.

We witness a good ending of this dangerous musical plot!

Stefan and the wolf (photo of the wolf by Retron, public domain)

Tone painting is a sub-discipline of "Onomatopoeia" which comprises processes of creating a word that phonetically imitates or resembles the sound that it describes. This can be found in many linguistic contexts. A frog croaking may be described differently in different languages. For animal sounds one could use quack for a duck, moo for a cow, bark for a dog, roar for a lion, purr for a cat, or baa for sheep. The cock cries out in German with kikeriki, in Dutch kukeluku, in French cocorico, in Spanish quiquiriqui, in English cock-a-doodle-doo, and in Japanese kokekokkoo. And there are many more examples.

Noteworthy is the element of tone painting for literature, e.g. in many poems. It is realized by combining several words. Here is one: The German poet Clemens Brentano (1778–1842) wrote a Lullaby in using words expressing sounds:
Singt ein Lied so süßgelinde,
Wie die Quellen auf den Kieseln,
Wie die Bienen um die Linde
Summen, murmeln, flüstern, rieseln.
The last verse could be translated as:
Hum, murmur, whisper, tickle.

References

1. W. Bühler, *Aristoxenos und Pythagoras - Ein elementarmathematischer Streifzug durch die Geschichte der musikalischen Skaken und Intervalle* (Peter Lang, Frankfurt a. M., 2017)
2. A. Baker (ed.), *Greek Musical Writings: Vol. 2, Harmonic and Acoustic Theory*, (Cambridge University Press, Cambridge, 1990), pp. 190-208
3. D. Benson, *Music: A Mathematical Offering* (Cambridge University Press, Cambridge, 2006)
4. W.A. Sethares, *Tuning, Timbre, Spectrum, Scale* (Springer, Heidelberg, 1998)
5. W. Odington, *De Speculatione Mucices* (first published in complete form in Edmond de Coussemaker's Scriptores de musica medii aevi (4 delen)) (1864–1876)
6. L Fogliano, *Musica Theorica Docte Simul ac Dilucide Pertractata: In qua Quamplures de Harmonicis Intervallis: Non Prius Tentatae Continentur Speculationes*, (G.A. Nicolini da Sabbio, Venezia 1529)
7. D. Bartel, Andreas Werckmeister's final tuning: the path to equal temperament. Early Music **43**, 503–512 (2015)
8. F.A. Kuttner, Prince Chu Tsai-Yü's life and work: a re-evaluation of his contribution to equal temperament theory. Ethnomusicology **19**, 163–206 (1975)
9. V. Galilei, *Fronimo Dialogo di Vincentio Galilei Nobile Fiorentino, Sopra L'arte del Bene Intavolare, et Rettamente Sonare la Musica*, (G. Scotto, Venice, 1584) pp. 80–89
10. H. Helmholtz, *On the Sensations of Tone as a Physiological Basis for the Theory of Music*, 2. edn. translated by A.J. Ellis (Longmans Green, Harlow, 1885)
11. L.S. Penrose, R. Penrose, Impossible objects: a special type of illusion. British J. Psychol. **49**, 31–33 (1958)
12. https://en.wikipedia.org/wiki/Harmonic_series_(music)

5

Comparison with Non-European Systems

5.1 Introduction

We will now try to address "non-European" music. But what is "non-European" music? Or ask the other way around, what is the "European" music? To answer this question, we have to start with some human history.

About 7 million years ago humans started to spread around the world out of Africa. Humans arrived in the southern part of China through the fertile crescent and in India at about 1 million and in Europe by 500,000 years BC. The former group reached Siberia by 20,000 BC and spread further to North America around 12,000 BC. Humans arrived in Australia from 40,000 BC on, and then in New Guinea (by 33,000 BC). Much later, from 3500 BC on, humans spread via Taiwan to the Pacific islands, Hawaii, Madagascar and New Zealand till 500–1000 AD [1]. The spreading of cultures including music followed accordingly. Music belongs to the human culture from very early times, as is suggested by the famous flute from the time of the Cro-Magnon men (see Fig. 1.2) [2].

The cultural evolution of music, melodies or songs during the ages often shows common features. Music initially happened spontaneously, when some people tried to do something together, for example, lifting a heavy stone. They had to call out together with precise timing. When they had to carry this heavy stone, they needed to move their hands and legs with regular rhythms synchronously. So they started singing. Moreover, when they were happy, they sang and danced together. When somebody died, they sang for the soul of the dead person (more in Chap. 14).

© Springer Nature Switzerland AG 2021
K. Tsuji et al., *Physics and Music*,
https://doi.org/10.1007/978-3-030-68676-5_5

Another common aspect is the concept of the "octave". Either in European or in non-European music, the octave existed at any time and anywhere. An octave is a tone interval in which the higher tone has twice the frequency of the lower one. We cannot change this physical fact. Moreover, the human ear perceives some octaves almost as a single tone. It means that octaves are not uncomfortable for human ears. We cannot change this physiological fact.

For expressing their emotions, however, humans used certainly some more intervals than octaves. If so, which intervals within an octave were used?

All over the world, one finds a preference for certain musical scales. Five notes per ascending octave (or six if the upper octave is also counted) and the corresponding way downward are widely used. This is now called pentatonics (see Chap. 3). And then, on the basis of pentatonic scales, the heptatonic scale with seven steps (usually five steps from a pentatonic scale plus two more tones) was born.

Around 500 BC in Greece, Pythagoras and some other philosophers started trying to understand music in a logical way [3], using a monochord to determine intervals[1] (see Chap. 3). Further on, church people studied and constructed their music according to the rules of church modes in the times of CharleMagne. There followed well-tempered tuning approaches up to modern equal temperament (Chap. 4). Probably this is the base of the European music. Interestingly, the Chinese developed their own musical theory, which is principally very similar to the Greek one, but developed further in a different manner.

In other areas there are also pentatonic and heptatonic scales, and even different kinds of scales. Such music remained intact at least until the 1800s. Such music we call "non-European".

The regions of non-European music can be, according to the human spreading process [1], divided into 6 areas as shown in Fig. 5.1: the Southern neighborhoods of Europe, its Northern neighborhoods, East Asia, the islands in the Pacific ocean (Austronesia) including Australia and New Guinea, Sub-Saharan Africa and the Americas.

Rhythms are also different from area to area. Not all music follows the regular sequence of "one two, one two", or "one two three four, one two three four". In some music cultures the metronome can not be applied. Also the concept of a "bar" as a rhythmic unit does not exist in some non-European music.

---

[1]It is not sure whether Pythagoras really used a monochord for his theory. The mathematician Euklid made mention of the monochord in 300 BC.

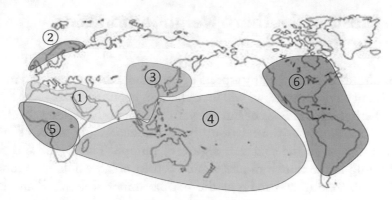

Fig. 5.1 Non-European music regions. 1: Southern neighborhoods of Europe, 2: Northern neighborhoods of Europe, 3: East Asia, 4: Austronesia, Australia and New Guinea, 5: Sub-Saharan Africa, 6: North and South America

Thus, we also refer to music from various countries using a quite different cultural way to develop modern sound.

Furthermore, a remarkable difference between European and non-European music is: European music uses chords and harmony a lot (for example, polyphonic instruments like the piano or the guitar), while in the non-European music mostly solo melody is accompanied by percussions or interlude with voice. Accompanying sound/noise contributes to make solo music more festive, more alive or thicker, depending on the solo melody.

What is the reason for such differences? It could be because of the development of musical instruments, the structure of societies, or the mentality of the people. Some of these considerations will be discussed in Part VI.

Throughout this chapter we use the brackets shown below to indicate the type of interval between subsequent tones:

$\wedge$: minor second (1 half-tone interval)

no sign: major second (1 whole-tone interval)

⌐⌐ : minor third (1 ½ whole-tone interval)

⌐⌐ : major third (2 whole-tone intervals)

5.2   Music in Southern Neighborhoods of Europe

Having already presented many aspects of pentatonics in our own and in some quite distant environments, we will focus here on heptatonic music, because in the southern neighborhoods of Europe various heptatonic scales were created.

Figure 5.2 illustrates the relationship of the heptatonic scales of church modes and those in the southern region of Europe. The minor and major scales belong to the church modes, which consist of 5 x whole-tone intervals and 2 x half-tone intervals. From the natural minor scale (Aeolian church mode) the harmonic minor was derived. It has 3 x whole-tone intervals, 3 x half-tone intervals and 1 x $1\frac{1}{2}$-tone interval (see Sect. 3.2.4). The Phrygian dominant scale was derived from the Phrygian mode and has also 3 x whole-tone intervals, 3 x half-tone intervals and 1 x $1\frac{1}{2}$-tone interval.

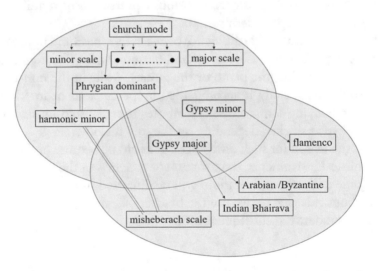

Fig. 5.2 Relationship of musical scales between European church modes and neighborhood music

5.2.1   Sinti and Roma Music

As a boy SCM remembers how much he enjoyed the music played by Romani people, who had visited the village where he was growing up. These colorfully dressed people mediated a longing for distant unknown worlds. Much later, he enjoyed in a similar fashion the music played by a group of Romani instrumentalists in a restaurant in Budapest—the music seemed to be of a similar kind, presented in a breathtaking "ferocity" and abundant musicality. So many times people have enjoyed the folk music of the countries where the Romani went through or settled. Frequently they formed bands with two violins, a cimbalom[2] and a double bass. The tours of "rajko"-orchestras—featuring young Romani-musicians—became popular. The boys were both endearing and virtuoso.

Romani music (which many still call Gypsy music) is wide-spread, not only in European countries, but also used to be quite popular in northern India, Iran or Jewish culture. In fact, originating from India these nomadic people have roamed through most European countries over the centuries.

With their creative musicality they have conquered or adapted to the tunes of many European countries, playing the popular Csárdás in Hungary, adding their flavor to the traditional Andalusian flamenco or applying their style to improvising the fasil music of Turkey, Transilvanian events in Romania, music genres in Bulgaria and Serbia. Thereby one cannot speak of "the Gypsy music", because the Romani, while acting as entertainers and tradesmen traveling wide distances, a unified Romani musical style did not develop, but it experienced a multitude of influences in melody, harmony, rhythm and formal structures from region to region.

Certainly Sinti and Roma follow quite special melodies and harmonies. Some of their characteristic musical scales include, beside a number of half-tone steps, also several unusual minor third steps. Typical scales are:

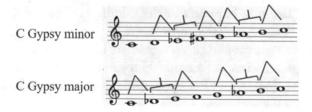

---

[2]A cimbalom is a chordophone composed of a large, trapezoidal box with metal strings stretched across its top.

The Gypsy minor scale (written above as an exampe with C4 as fundamental), also known as Hungarian Gypsy scale introduces to harmonic minor a second hiatus (augmented second) between the 3rd and 4th step, thus generating a special oriental sound.

In a variant called Gypsy major (or Byzantine or Arabic scale) the 2nd note of a harmonic minor scale is lowered by a half tone, and there is a hiatus between the 2nd and 3rd step. It can be considered as a plagal form of the Gypsy minor scale starting on the 5th step. One can also consider the Gypsy major scale by starting from C major and diminishing steps 2 and 6 of the scale by a half tone. Interestingly, the first 6 tones of the Gypsy major scale are exactly the same as those of the Phrygian dominant: the 7th tone of the Gypsy major scale is a half tone higher than that of the Phrygian dominant (see Fig. 5.3).

Since Sinti and Roma people live almost everywhere in Europe, their music is familiar to European ears. European composers used Gypsy scales in many pieces. In this sense the Sinti and Roma music stands on the border between European music and non-European music.

A good example is Geoges Bizet's use of the above scales in his opera "Carmen".

Fate motif of the opera Carmen, Georges Bizet

The famous fate motif appears in the middle part of the overture and is repeated many times during the entire opera.

The Hungarian pianist and composer Franz Liszt collected plenty of Hungarian folksongs and composed many Hungarian Rhapsodies. In the Rhapsody No. 2 for piano he uses, after a powerful Lento-Introduction, an Andante movement in C♯ Gypsy minor.

Hungarian Rhapsody No. 2 of Franz Liszt

5.2.2 Misheberach Scale

There is an interesting heptatonic scale, the misheberach scale, frequently used at the beginning of a Jewish prayer ("May he who blessed"), and presented on Sabbat morning in the synagogue.

This scale is formed starting from the Dorian church mode (on D) and augmenting the 4th step from the 4th to the tritone. Similar to Phrygian dominant and harmonic minor it has also 3 × whole-tone intervals, 3 × half-tone intervals and 1 × 1$\frac{1}{2}$-tone interval (see Fig. 5.3).

The Phrygian dominant scale starts with tone E, and is also called the Oriental, Spanish or Jewish scale. The 3rd note is augmented by a half tone, such that the original minor third is transformed to a major third. This scale is quite popular in Spanish flamenco or in various kinds of Jewish music. It is used also in Indian and Arabic music, as well as in Hebrew prayers. The flattened 2nd and the augmented step between the 2nd and 3rd tones create its distinctive sound.

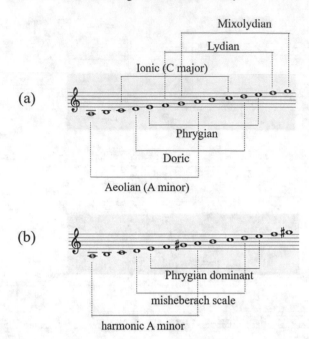

Fig. 5.3 Comparison of church modes (a) and misheberach relatives (b)

In addition, we have the "Ukrainian Dorian scale", which is quite similar to the misheberach scale as used in Jewish music, also played extensively in Romanian dances and mazurkas and thus figuring in Eastern European music full of exotic and romantic flavor.

5.2.3 Ragas in Indian Music

Different from the European music, in Indian music melodies are roaming more freely in pitch. The option is freedom of melody, and accompanying harmony must be much simpler.

An important form of Indian music are the "raga", meaning in Sanskrit mood or color or dyeing. These are improvised pieces of music with motifs having the ability to "color the mind" and affect the emotions of the audience. The dominating instrument for many ragas is the sitar, as depicted in Fig. 5.4. Ragas range from small and short songs to big songs which have great scope for improvisation and for which performances can last over an hour. Each raga has traditionally an emotional significance and symbolic associations such as with season, time and mood [4]. A description of ragas would fill thick chapters and no corresponding attempt can be made here.

Fig. 5.4 Sitar (Photo by KT in Volker Bley's gallery)

This classical format tends to use 1 or 2 melody instruments (the sitar and a kind of flute) with percussion (bongos). It starts from a heptatonic scale with a non-equal tempered system. This allows more room for expressive playing in the melody by swooping up and down around the 7 notes of the scale. Each note will eventually turn out to be the one which fits in with the harmony of the accompanying chords. For an Indian instrumental musician there are many more options available as a desired pitch—which is why it would be so difficult to prepare an appropriate harmony. However, over the past 50 years, much of the popular musical output from India has followed the European pattern and uses the equal temperament.

Ragas are associated with a fixed tonal structure by prescribing which tones fit to a piece of music, including melodic and ornamental elements to be played with certain tones. Scales are based on seven main tones, the "Svaras". The scale starts with "Sa", often considered as the fundamental tone (like the "C" in our scales), but is then also quite dependent on the particular raga having higher or lower pitch.

Seven "degrees" of tones in a musical scale are shared by many raga systems and put them close to the European heptatonics. Quite remarkable is the concept that the octave has 22 "shrutis" or microintervals adding up to 1200 Cents. There are intricate lists assigning specific variations in pitch to these microtones. The system with each shruti calculated to 54.5 Cent is compared to the ancient Greek one with an enharmonic quarter tone calculated to

55 Cent. This forms a rather vague basis for some music theories employed in some scale applications (compare "microtones" in Sect. 3.2.6 of Chap. 3).

There is no precise notation rule in Indian music. The tones are documented by letters, underlignments or postposed numbers.

One has developed, at the beginning of the 20th century, a binding scale system relating Indian ragas (restricted to 10 characteristic types of melodies) with heptatonic tone sequences being transcribed to European-type heptatonic scales. One finds that seven of them are identical to scales of church modes, one corresponds to the Gypsy scale and two do not have any corresponding scales. They are Kalyāna (= Lydian), Bilāvala (= Ionic), Khamāja (= Mixolydian), Mārava (= Lydian b9), Kāphī (= Doric), Āsāvarī (= Aeolian), Bhairavī (= Phrygian), Bhairava (= major Gypsy), Pūrvī (no corresponding scale) and Todī (no corresponding scale).

According to the phase portrait which we have developed, we can have a glance which scales are relatives of European scales and which are different (Chap. 13).

5.2.4   Arabic Music

A large cultural domain, for which musical developments have played a major role, is found in Arabian countries, the classical and popular music forms of the people of the Arabic world (the Magreb states, Egypt, Jordan, Syria, Lebanon, Iraq, and so many others). From a European perspective, Indian and Arabic musical tunes and pieces share certain common features, although their characteristic sound differs distinctly. Thus, it is not so difficult to construct a few floating bridges between both music styles.

Arabic music, not unlike Indian music, obtains its expressive effects mainly by its melodic and rhythmic impact, less through harmonic means. In fact, a large portion of Arabic music has traditionally a linear unisono structure. There are some genres of Arabic music that are polyphonic, but typical Arabic music is homophonic.[3]

Small ensembles are common, which are structured in a similar way over the entire Arabic cultural domain. Among the instruments the arabic lute (oud) has a leading role, followed by a number of flutes and then by drums. Certain instruments belong to the folk culture, like the guitar (qitara), the rebab with just one string and a small corpus such as half of a coconut (related to the small

---

[3] Heterophony presents a dominant form for musical performance, but different from mono- or polyphony in European music: All play the same melody but add some improvisation and ornamentation. The role of singers is very strong, since the compositions allow for great variation in interpretation and improvisation.

ancestor of the violin, the rebec). These instruments appear in the painting shown in Fig. 5.5.

Fig. 5.5 Two sections from a painting of a Persian family playing music by Ibrahim Jabbar-Beik (1923–2002), among various instruments, some drums, a rebab (left), a special oud, a flute, and a dulcimer (right, also refer Fig. 10.1) are seen. (public domain, photo by Nightryder84)

There exists a rather strict tonal system with its own interval structure, including microtones, not so different from the Indian raga scale system (invented already in the 10th century). This is called the maqām system, which is also common in Iran and Turkey. Maqāms can be realized with either vocal or instrumental music without a rhythmic component.

A maqām usually covers only one octave. An example of the tone level is shown below.

A typical tone level of maqāms

There are about 100 maqāms, which can be divided into the four groups listed in Table 5.1. One finds here beyond the accidentals for half tones also those for microtones, indicated by special signs in note script and meaning that the corresponding note is diminished by a quarter step, i.e. there is a $\frac{3}{4}$

Table 5.1 Examples of Maqāms

Scale	1st	2nd	3rd	4th	5th	6th	7th
Rāst	C	D	$\frac{3}{4}$E	F	G	A	$\frac{3}{4}$B
Nahāwand	C	D	E♭	F	G	A♭	B
Hiğāz	D	E♭⁺	F♯⁻	G	A	B♭	C
Sīkāh	$\frac{3}{4}$E	F	G	A♭	B	C	D

tone between this and the previous note. (for more details of microtones, see Sect. 3.2.6).

Introducing such quarter tones was a compromise in the Arabic music world, due to the fact that in different regions one used differing tonal accentuations. For instance, in the maqām Hiğāz the 2nd tone is slightly augmented (E♭⁺) while the 3rd tone is somewhat diminished (F♯⁻).

In 1932, at the Congress of Arab Music held in Cairo, Egypt (attended by Béla Bartók and Henry George Farmer), experiments were done that determined conclusively that the notes in actual use differ substantially from an even-tempered 24-tone scale. Furthermore, the intonation of many of those notes differ slightly from region to region (Egypt, Turkey, Syria, Iraq).

Both in modern practice, and evident in recorded music over the course of the last century, several differently-tuned E's in between the E♭ and E♮ of the European Chromatic scale are used, that vary according to the types of maqāms, and the region in which they are used. Arabic music does not modulate to 12 different tonic areas like the Well-Tempered Klavier. The most commonly used quarter tones are on E (between E and E♭), A, B, D, F (between F and F♯), and C.

A direct comparison between maqāms and church modes is provided in Chap. 13 by using phase portraits. For instance, changes in the basically heptatonic tonality occur due to the effect of microtones.

5.3   Music in the Northern Neighborhoods of Europe

5.3.1   The Sami

On our journey around the world we make a stop in the north of Europe and visit the Sami who have dwelled in a vast region in the north of Scandinavia for many centuries, sharing their living space with the nations of Sweden, Norway,

Finland, Russia (and with reindeers). In fact, they were virtually unknown in Roman times, but then cultivated a close relationship with their Scandinavian neighbors. They developed their own way of worshiping deities in the frame of their beautiful nature under preservation of many ethnic traditions (see Fig. 5.6).

Fig. 5.6 A Sami vocalist performing at the 2007 Riddu Riddu festival (CC BY-SA 3.0 by Mates)

A traditional form of music is the "joik", similar to yodeling, an archaic form of singing [5]. This quite monotonous and guttural singing of the Sami emphasized words more than music. They are dealing with persons, animals and natural phenomena. This way, the Sami feel closer to persons of their choice. Sacred places in nature, feelings and hopes are well-known themes of the joik. The traditional and most popular ones have as their theme the wolf and the reindeer, and personal ones describe a particular character. The only instrument for accompaniment was the fadno, a kind of flute made from the stems of single reeds. In spiritual sessions, when joiks were sung, one also had the Schaman beating his magic drum. He predicted future events according to the path which a piece of wood took through the symbols on the drum's membrane.

From the 18th until the 20th century the joik was forbidden in the course of christianization as expression of the old religion and its performance even punished. Only after 1870 there occurred a revival of this deep symbol of Sami culture. We can enjoy the joik singing in Fig. 5.6.

5.3.2   Celtic Music

Many people and ethnicities have left their particular traces in history radiating to our modern times. The Celts belong to an outstanding group of musical "strongholds". Modern music from the Celts originates mostly from Scotland, Ireland or Wales. In our actual realm of pop, jazz, and many forms of popular music we normally like to listen to this Celtic music, which has often adapted to the reception of familiar sound and performance.

There are various types of Celtic musics. Commonly they have a melody, which moves up and down the primary chords. Melodies can be varied by a harp or pipe. In some songs the tonal interval is relatively wide so that the poetic line can be stressed. The notes go with a continuous melodic flow, without harmony or chords. In this sense it has a solo musical form: one may hear the sound of several solo parts sung or played simultaneously. The melodies are mostly pentatonic in nature.

Probably therefore, when it comes to listening to more traditional tunes and instrumental accompaniment, we also may notice some unfamiliarity, attractive or not, which documents much of the special origins of Celtic culture and music. And we may ask the question, whether or not Celtic tonal "products" just belong to what we often name "European" music, or how much they carry along their individual character that has been maintained over the centuries of their migrations.

The people of the Celts walked a long way. From 800 BC on until times of Roman dominance this people spread from southern Germany westward into modern France, southwest into Iberia, southward into northern Italy, and even migrated over the Balkans to Asia Minor. After Cesar's victory over the Gauls, they retreated to north-western regions of Europe, especially the British islands, and left their social and cultural footprints until today in Ireland, Scotland, Wales, Cornwall, Brittany, and on the island of Man.

In their powerful times of wealth and dominance the Celts developed various kinds of brass instruments, many of them being useful during warfare. An outstanding exemplary is the carnyx (known since 300 BC), a variant of the Etruscan-Roman lituus. It consists of a tube between 1 and 2 m in length, ending in a bell constructed as an animal's head (Fig. 5.7a). The sound of a carnyx was harsh.

Around the same time there were two other brass instruments played by Celtic warriors: the Celtic horn (Fig. 5.7b) and the Celtic trumpet [6]. The horn was a large oval-curved horn with a thin tube and a modestly large bell and had a cross-bar as a means of supporting the instrument's weight on the

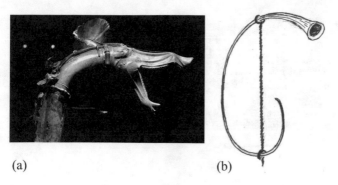

(a)   (b)

Fig. 5.7 a A wild boar, the head of a carnyx (CC BY-SA 3.0 by Claude Valette), **b** a Celtic horn (by KT)

player's shoulder. It is also most likely of Etruscan origin. The trumpet was similar to the Roman tuba and came in different lengths.

It is not clear how such brass instruments and tones played with them influenced modern Celtic music, which is familiar to us. We do not know what kind of music may have been played in ancient times, but we can imagine that its essence has inspired modern Celtic music.

When listening to Arabic or Middle Eastern music one finds common characteristics similar to the traditional music of Scotland and Ireland. In both cases there is a strong emphasis on melody and rhythm. Here we see some connection between Celtic and Arabic music, which occurred when the ancient Celtic people spread.

5.4   Music in East Asia

5.4.1   Chinese Music

Not only in the fertile crescent but also in China there is archaeological evidence for musical culture: bone flutes from 8000 years ago have been found in Jiahu village.

Much later, during the Zhou dynasty (11th—3rd century BC) a twelve-pitch (chromatic) scale was established. This has the same intervals as the Pythagorean scale, based on ratios of 1:2, 2:3, 3:4, 4:5, etc. On the basis of this twelve-pitch system 84 scales were created: 12 half tone x 7 different scale groups. These pitches were "frozen in" by bronze bells [7].

For easier understanding we symbolize the Chinese tone system as a kingdom. In this kingdom there are 12 clans. Each clan has a boss. The name of

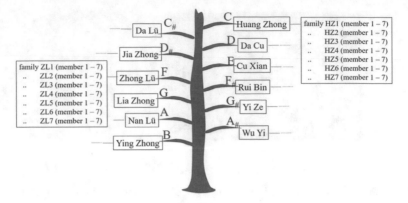

Fig. 5.8 Chinese tone families

Table 5.2 Example of 7 scales of the clans Huan Zhong and Zhong Lü

	1st	2nd	3rd	4th	5th	6th	7th
Huan Zhong 1	C(boss)	D	E	F♯	G	A	B
Huan Zhong 2	D	E	F♯	G	A	B	C
Huan Zhong 3	E	F♯	G	A	B	C	D
Huan Zhong 4	F♯	G	A	B	C	D	E
Huan Zhong 5	G	A	B	C	D	E	F♯
Huan Zhong 6	A	B	C	D	E	F♯	G
Huan Zhong 7	B	C	D	E	F♯	G	A
Zhong Lü 1	F(boss)	G	A	B	C	D	E
Zhong Lü 2	G	A	B	C	D	E	F
Zhong Lü 3	A	B	C	D	E	F	G
Zhong Lü 4	B	C	D	E	F	G	A
Zhong Lü 5	C	D	E	F	G	A	B
Zhong Lü 6	D	E	F	G	A	B	C
Zhong Lü 7	E	F	G	A	B	C	D

the boss is C for clan Huang Zong, C♯ is for clan Da Lü, and so on. And each clan consists of 7 families. Figure 5.8 illustrates this organization as a family tree.

Let us look into a family. As examples we go to the clans Huan Zhong and Zhong Lü, the clan boss is C and F, respectively. The families and their members are listed in Table 5.2. (Original Chinese notations are translated into English notations.)

Looking more closely, one notices that the 5th scale is exactly the E major in the Huan Zhong clan and C major in the Zhong Lü clan (the Ionic). On comparison with the church modes (see Fig. 3.2), the scales in the table correspond to 1 = Lydian, 2 = Mixolydian, 3 = Aeolian, 4 = Locrian, 5 = Ionic, 6 = Dorian and 7 = Phrygian on E for Huan Zhong and C for Zhong Lü.

Different from the European chromatic scale, the Chinese twelve-pitch scale was not used directly for their music. From this set of notes they constructed other scales, namely pentatonic scales. Their five tones correspond to the five elements of nature: water, fire, earth, trees and metal. (A theory of the cycle of fifths (see Sect. 3.2.4) was also developed during the Zhou dynasty.)

While European music developed in a church environment, Chinese music was established as ceremonial music for the royal court. During the Tang dynasty (618–907 AD), due to Confucius pentatonic music bloomed. Especially, for a kind of Chinese harp "qin", many pentatonic pieces were composed.

Of course, music was played not only for ceremonial purposes in court but also for fun among people. Chinese music, both court and folk music, were influenced by inner Asia and India via the silk road, as well as by the Mongols. They brought along new instruments and different systems [8]. But essentially it had pentatonic melodies accompanied by percussion.

European music had no chance to penetrate into China till the 19th century. Also there was a big break during the cultural revolution (1966–1976). Nowadays, however, a lot of Chinese students study European classic music and often win international competitions.

Figure 5.9 shows some basic pentatonic scales. Figure 5.10 is an example of the Chinese notation (Guqin notation) of a song "Elegant Orchid" from the 6–7 centuries AD.

5.4.2 Japanese Pentatonics

Since the 7th century Japan sent missions to China (under the Sui and Tang dynasty) to learn about highly developed cultures. Through these missions, Chinese pentatonic scales (see Fig. 5.9) were brought to Japan during the Heian period (794–1185). As usual like nowadays, Japanese liked to modify what they learned to fit to Japanese life, and soon after they developed their own Japanese scales (see Fig. 5.11) [9].

The Ritsu scale served mainly for the ancient Japanese court music (in Japanese: gagaku), and it is played even now in Shinto shrines. Music for city people was played according to the "In" or "Miyako" scale used often for the Japanese traditional string instruments, Koto and Shamisen. Similar to the harmonic minor scale, the Miyako scale has different ways for ascending and descending cases.

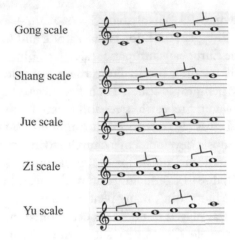

Fig. 5.9 Chinese pentatonic scales

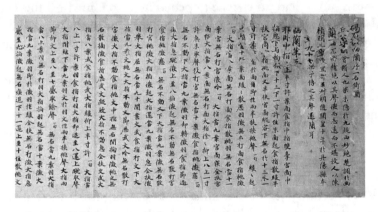

Fig. 5.10 The earliest music notation discovered is a Guqin music, "Elegant Orchid" in the Jieshi Tuning (Jieshidiao Youlan) (public domain)

A famous example of the Miyako scale is the "Cherry Blossom". Here, the fundamental tone is E.

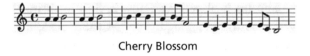

Cherry Blossom

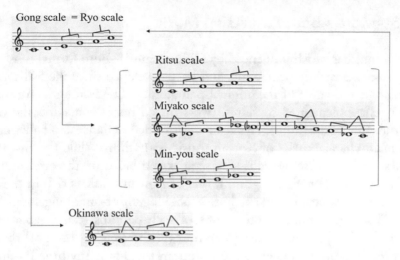

Fig. 5.11 Development of Japanese scales

The Min-yo scale is used for down to the earth purposes in the countryside: Japanese traditional folk songs and children's songs.

A special case is the music on the Okinawa islands in the very south of Japan. The rhythm is also different from the typical Japanese regular one, as shown in the song "Tanchame nu hama" (an Okinawa folksong).

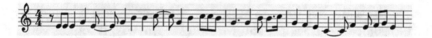

Tanchame nu hama

An interesting fate can be told about the "Ryo" scale. It had practically disappeared from Japan since "Ritsu" became standard. But during the Meiji restauration in the 19th century, the European system of pentatonics of the major variant was reinstalled. It means that the "Ryo" revived. Also a minor-mutant from the European minor-variant became common in this country, as of today.

In many Japanese pentatonic music pieces, even if starting with the notes C or E, they end remarkably with notes A or D.

5.4.3 More About East Asian Music

Korean music is much influenced by Chinese music. Both formal music and folk music are mainly pentatonic, and have developed since the 5th century. Hyangak is the name of the original pentatonics. The leading instrument is the oboe, supplemented by flutes, two-stringed instruments and the chang ko-drum (hourglass drum). Hyangak is performed in dance and ballet music.

A main characteristics of Korean music is rhythm. While the rhythm of Chinese and Japanese music is very regular and based on the even numbers (two-two, four-four, eight-eight,...), that of Korean music is rich in different patterns: short-long, long-short, short-short-long, short-long-long, long-short-long. There are various combinations of such patterns (called jangdan). It sounds for European ears like a three-four (or three-eight, six-eight) rhythm. However, it is not identical to the European three-four rhythm. Traditional Korean music does not have a concept of the "bar", which is a unit of a short cycle of rhythm in European music. Instead, Korean music has a longer cycle of rhythms, where the different patterns appear alternately. Moreover, vibrato and ornamental notes create a peculiar atmosphere. This type of rhythm can be called "fuzzy".

When we look further north, there is a special singing—overtone singing called "Tuvan throat singing"—sung in Mongolian, Tuvan and Siberian. Here harmonics is created by changing the shape of the resonance cavity of the mouth, larynx and pharynx, producing several pitches (fundamental and overtones) at the same time (refer to Sect. 8.3). To describe how it sounds is quite difficult: something cracking, and whistling.... Anyway, it sounds very unfamiliar to the people who have grown up in the environment of European music.

Such overtone singings exists in Tibet, Pakistan, Iran, Afghanistan and Japan, too.

Looking at South Asia, the characteristics of music comes close to that of Austronesian music (see Sect. 5.5.1). In Thai music one octave is (approximately) equally divided into 7 tones. However, the music itself is pentatonic. One unit of a piece possesses 5 tones, the next unit is modulated, resulting in the use of a different group of 5 tones. Subsequently, 7 equally divided tones are used not in a manner of a heptatonic scale, but a pentatonic scale [10]. Since Thailand is located close to the intersection of China and India, and also lies on the trade routes from west to east, various instruments of Persian origin and Indian origin are used. Each instrument has slightly different pitches, but it does not matter so much in Thai fuzzy music. The fuzziness is also common in Austronesian music.

Fuzziness:

According to the Cambridge dictionary, "fuzziness" is the quality of not being clear to see or hear, or the quality of being not detailed or exact enough. According to Thesaurus, "fuzziness" is the quality of being indistinct and without sharp outlines. In the fuzzy music either pitches are not exact, or rhythms are not exact.

5.5   Music in Austronesia, Australia and New Guinea

Humans arrived in Australia about 40,000 BC, and then in New Guinea around 33,000 BC. There was a small land connection between the northern part of Australia and New Guinea. Otherwise they were completely isolated.

Much later ($\approx$ 3500 BC) the so-called "Ta-p'en-k'eng culture"[4] from south China reached Taiwan and through there further on many of the widespread islands [1]: Easter Island, Hawaii, Madagascar,...The native Taiwanese are called Taiwanese Aborigines, and their languages (Austronesian language family) influenced such islands (except New Guinea and Australia). We suppose that the musical culture of these regions is also related to their language development.

5.5.1   Austronesia

As mentioned above, about 3500 BC the ancestors of Taiwanese aborigines arrived on the Taiwanese island. Some of them moved further to many Pacific islands. Others stayed in Taiwan and developed their own aboriginal culture, until the island was colonized by the Dutch in the 17th century. Their culture, including languages, almost disappeared during various colonial periods, and finally with the arrival of the Nationalist Party of China. In 1987 there started a revival of traditional culture. However, modern Taiwanese culture is not any more the original one. Their original music is pentatonic and polyphonic, similar to other Pacific islands.

---

[4]A distinctive decorated pottery of South China

Gamelan

Gamelan is a famous form of Indonesian music, an ensemble of tuned percussion instruments [11]. It is regionally diverse in size, tunings, and instruments. It consists of three main instrumental groups. These are (1) tuned percussion-like metallophones for melody, (2) tuned percussion-like large kettle gongs and hanging gongs and (3) untuned percussion-like drums for rhythm (see Fig. 5.12). Each group of percussion instruments creates different rhythm: one per bar, two per bar, four per bar, one in four bars, one in 8 bars, and so on. The cycle is longer for the lower tone instruments. Similar ensembles are also found in Malaysia.

Fig. 5.12 A gamelan musician playing bonang. Gamelan Yogyakarta style during a Javanese wedding (CC BY-SA 3.0 by Gunawan Kartapranata)

"Slendro" (pentatonic) and "pelog" (hexatonic) are the two scales often used for the gamelan. Strictly speaking, it is not possible to write down their scales on the European five-line system. First of all, there are no exact pitches. They differ from region to region, and from ensemble to ensemble and even from instrument to instrument in one ensemble. The next and more essential fact is that some pitch differences to the next tone correspond neither to a whole tone nor to a half tone (even not to a 1/4 tone). This concept of music is completely different from that of European music, where exact pitches and harmonies are important. Here we also find a world of fuzzy physics.

Nevertheless, to get some ideas of their music we show approximated notations of slendro and pelog:

slendro   pelog

Melanesia, Micronesia, Polynesia

Most kinds of the traditional Austronesian music have become extinct or strongly changed, since Europeans have arrived. However, some of them have survived and were revived.

Vocal music is a common feature of the oceanic music. Most of the songs are monophonic and strongly connected to dances with body percussions. Music is inevitable for religious ceremonies, workings, battling and the whole lives of the inhabitants. There are songs only for men, and also songs only for women [12].

In Melanesia one has not only monophonic songs but also chorus with up to five parts. Often songs are sung in a form of response between a leader and the chorus ("call and response"). The "call and response" style is seen in music in sub-Saharan Africa (Sect. 5.6), as well as in Mesoamerica (Sect. 5.7.3). In most of the cases very sophisticated instruments are played by only adult men. A slit gong is one of the famous instruments.

In Micronesia instruments are not so important as songs. Often one singer performs, dances, and also plays rhythmic instruments like percussive sticks. Movements of hands and arms are important.

Among Polynesian islands New Zealand and the Chatham islands are the youngest places for the Austronesian expansion. They arrived in New Zealand around 1000 AD and in the Chathams around 1300 AD. The traditional music of the Māori in New Zealand is microtonal in a single melodic line, which is centered around only one note. One believed that something bad would happen if a song is interrupted. Therefore, a solo singer sings a very long phrase in one gulp, or in a chorus singers breath not at the same time but group-wise in different timing so that a song is sung without interruption.

Dance traditions are essential for Polynesia. For example, Tongan, Tahitian and Hawaiian dances are famous.

Fig. 5.13 An Indigenous Australian man playing a didgeridoo (CC BY 2.0 by Steve Evans from Citizen of the World)

Surprisingly, in Madagascar, too, Austronesian ethnic groups have lived since 500 AD. The traditional Malagasy music reflects the Austronesian culture, especially in the highlands. However, different from the vocal-based Austronesian music, a more polyphonic music from the African mainland prevails.

5.5.2   Australia and New Guinea

Aboriginal Australians have a unique culture since 40,000 years. Among their musical instruments a didgeridoo is one of the oldest instruments in world history. It is a kind of aerophone of a long tube without finger holes (see Fig. 5.13). The wood of Eucalyptus trees is used for the tube.

For these Australians music connects strongly to their daily life. There are songs for clans, for ceremonies and funerals. Noteworthy are their songlines (called Yiri), which function as a navigator. Such storytelling songs describe how the lands were created and named in the early times [13]. When the songs are sung in the appropriate order, they can travel the deserts on long distances.

Similar to aboriginal Australians indigenous New Guineans have their own culture since 35,000 BC without external influence. The outside world knew very little about their traditional music. Their traditions and cultures were disturbed by missionaries from the 1870s on.

The traditional celebrations called "sing-sing" include song and dance. During dancing a leader and a chorus sing the same song with some time delay like a canon (see Sect. 9.3.1). The percussionist Mickey Hart was the first person who commercially released the traditional music to an international audience in 1991.

5.6   Music in Sub-Saharan Africa

African pre-colonial music can be divided into three regions: northern Africa, the Sub-Saharan part and eastern Africa including Madagascar. The music of northern Africa is of Arabic origin, and we have described this in Sect. 5.2. The music for Madagascar is included in Sect. 5.5. In this section we introduce the African music of the southern part of the Sahara. Of course, this sub-Saharan area is huge, and each ethnic group has its own music. Let us summarize some common aspects of the music in this area [14].

Traditionally, African music is a kind of medium to express "life", and is usually played with dancing. Music accompanies anytime, anywhere: on the occasion of child birth, marriage, funerals, among others. It could be used for communication, too. Therefore, the rhythms of drums are important. If we categorize music sketchily: European music is specialized in harmony, Asian music in melody and African music in rhythms.

KT used to do aerobics with a trainer from Togo (Togolese Republic). It was a little difficult to dance synchronously with him, but somehow it was nice and she enjoyed it very much. One day he was ill and another trainer instructed the group: one, two, three, four, one, two, three, four,.... No problem with synchronization, but she felt that everything was too square. The rhythms of the African trainer often slightly deviated from one, two, three, four, one, two, three, four by maybe a 10th of a second or even less. Such irregularity can make dance smoother and rounder.

In this sense, it is almost impossible to write down their rhythms on the normal music sheets. African rhythms are polyrhythmic, and maybe more complicated than playing the piano: 16th notes with one's left hand, and triplets with the right hand. Many groups play different rhythms with syncopation and irregular rests.

Melody is coupled with such polyrhythms by hand clapping, foot stamping, and also dancing. Pitches are quite different from well-tempered scales and, therefore, to document exactly is difficult. The scale could be pentatonic or heptatonic. An octave is approximately equally (or unequally) divided into five or seven steps. Many melodies are very simple: a basic tone plus one or two neighbor tones. They are often constructed as a call by a vocalist or an instrument and its response by chorus or instruments, repeating short phrases several times. This "call and response" system has been exported to the American continent together with slave trades (see Sect. 5.7.3). A very simple example of a call-response song is shown below (Fig. 5.14).

Fig. 5.14 An example of call and response song

To explain (or to understand) their harmony is also difficult, because their concept of harmony is quite different from the European one. In African music it seems that the second tone above the main tone is their harmonic tone. For example, if the melody is based on the pentatonic scale as shown in Fig. 5.15a top, the harmonic tone of C4 is E4, that of D4 is G4 and so on. If the melody is based on a C major scale (Fig. 5.15b) then the harmonic tone of C4 is E4, that of D4 is F4 and so on. Note that these scales are not identical to the European scales, therefore, harmonic tones deviate from the tones shown here as examples. They are approximately minor and major thirds, and (perfect) fourths, and often two or three pitches are played simultaneously (polyphonic melody).

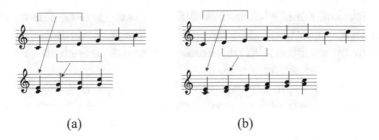

(a)   (b)

Fig. 5.15 African harmony for **a** pentatonic scale and **b** C major scale

In other words, maybe there exists no concept of harmony. The concept of a second tone can be also part of melody. It means that a melody "C4, D4, E4, G5, A5" is for them identical to the melody "E4, G4, A4, C5, D5". They have more freedom to choose tones as minor and major thirds, and (perfect) fourths for the melody. The melody of a response in a call-response song is also based on this rule (see Fig. 5.14).

For the instruments there are a lot of percussion instruments, which can be divided into instruments for rhythms and those for melodies. The former are various drums and shakers, while the latter are gongs, rattles and bells. There are also many melodic instruments: musical bows (Fig. 5.16), harps or harp-like instruments, krars and pluriarcs (Fig. 1.13), various types of xylophones and lamellophones, wind instruments like flutes and trumpets.

Such African music influenced the American music (salsa, samba, gospel, blues and others) in connection with the export of slaves.

Fig. 5.16 Obu man playing a musical bow, obubra, Cross River State, Nigeria (public domain)

5.7   Music in the Americas

5.7.1   History

Each continent in our world has its particular cultural traditions including music with its creation, thriving, and performance. When considering the American continent one certainly cannot expect, due to its comparatively late population, to find archaeological "miracles" as from the neolithical times or from Mesopotamia. However, there are many relevant discoveries in the context of "Making Music", which we will allude to in the following sections.

The first settlement of the American continent took place more than 15,000 years ago during the last glacial period across the Bering land bridge to Alaska and took place in several waves. More than 1000 years later the first groups of humans arrived at the southern tip of South America, in Terra del Fuego.

In the course of millennia numerous advanced civilizations came into being. Many of these high cultures perished, but some of their populations have survived, some of them to be named below (Sect. 5.7.3).

Our musical path does not follow directly the ways of human settlement from (probably) north to south, but considers more local or regional indigenous groups of people who have developed their individual culture and music, often much influenced by other colonizing or deported people who brought along their own habits and their way of life.

Indian music in Canada, north, middle and south America is characterized by magical and ritual-related songs. In their music a very highly and emphatically stressed voice is involved and this voice disappears very slowly and stepwise. At the beginning it is loud, then relaxed and with a quiet murmur. A soloist is accompanied by unisono chorus. The scale is mainly pentatonic.

The Czech composer Antonín Dvořák spent three years in the United States (1892–1895), where he wrote some of his most famous works. Among them the Symphony No. 9 ("From the New World") is the best known. The influence of the country on his music became evident by including Indian melodies, pentatonic scales, syncopations, and Scottish elements of rhythms. The second movement (Largo) of his 9th symphony he called "legend". It is a moving song of mourning based on a scene in a poem about the legendary and mythical forefather of the Iroquois tribe, Hiawatha, who laments for the death of his love Minnehaha. The melody of grief is sung by the English horn in sublime tranquility.

Dvořák also studied the music of "African-American spirituals" in which pentatonic elements prevail. Probably he obtained inspiration from one of the most famous songs, "Swing low, sweet Chariot". In this song you can hear a pentatonic scale, as well as syncopation in rhythm, which is typical for spirituals.

Swing low, Sweet Chariot

The spirituals are a distinct genre of music created by African Americans. They were inspired by Christian values while also describing the hardship of slavery. After being initially unison songs they are nowadays mostly known in choral arrangements. Spirituals are a musical form that is indigenous and specific to the religious experience in Africa. They have resulted from the interaction of music and religion from Africa with music and religion of European origin. Syncopation was a natural part of spiritual music. The rhythms of Protestant hymns were transformed and the songs were played on African instruments.

5.7.2   Blues

The term "blues scale" refers to several different scales with differing numbers of pitches and related characteristics. A main feature of the blues scale is the use of blue notes, that are sung or played at a slightly different pitch from standard for expressive purposes. The alteration is between a quarter tone and a half tone. (Note that such scales with alteration may not fit the traditional definition of a scale.)

The evolution of this scale may be traced back to Asia (pentatonic major) through native North America (pentatonic minor) with the addition of the ♭fifth blue note of African Americans [15].

There are some different scales of blues: the hexatonic, heptatonic and nonatonic blues.

The hexatonic, or six-note blues scale consists of the minor pentatonic scale plus the diminished fifth. This added note can be spelled as either a G♭ or a F♯, when starting with the tonic C (first note of the scale) as shown in the following example:

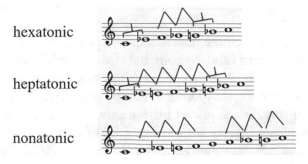

hexatonic

heptatonic

nonatonic

In the heptatonic blues scale shown below there exist a large and a small third as well as a pure and a diminished fifth. Blue notes are located between the small intervals formed by these two pairs of notes, thus do not exist in our chromatic tone system. One can approximate these notes by playing the adjacent tones together. But in fact, the dyad E-E♭, for instance, sounds quite different from a single tone between these two notes.

In a nonatonic version of the blues scale is a chromatic variation of the major scale featuring an E♭ and a B♭, which create a blues alteration. A non-formal way of playing this scale is the use of quarter tones. Guitar players manage this by raising a given note accordingly through bending.

These quite special notes, the blue notes, are usually lower than one might expect, comparable to a microtonal effect common in Oriental music, frequently involving a glide, either upward or downward. And one may speak of neutral intervals, neither major nor minor. Blue notes are often associated with a "blue" feeling in many folk songs. They are called "long notes" in Irish music.

5.7.3 Central America

An early cultural territory in Central America is called "Mesoamerica". It comprises large domains of nowadays states Mexico, Belize, Guatemala, El Salvador, Honduras, Nicaragua, and Costa Rica. In this wide area many pre-Columbian people lived, before Spanish colonization began. Here one finds advanced cultures like those of the Olmecs, Zapotecs, Maya and others.

There is no real evidence, however, to which extent music played a role in the cultural context of these people [16]. But we know from the wall paintings (dating from 775 AD at the Bonampak ceremonial complex near today's border with Guatemala) that Maya people played instruments such as trumpets, flutes, whistles, and drums at the occasion of celebrations, funerals, and other rituals (see Fig. 1.21).

As documented by a Spanish friar, Aztec music was a constant and important part of life, not only used for enjoyment, but also as a way of passing on culture. The educational system took care of teaching songs, often sacred hymns or "ghost songs" recounting the great deeds of the past. Drums played a big role in the music of the Aztecs, just to mention the teponaztli, a horizontal log drum, played with mallets (shown in Fig. 1.22 of Chap. 1). To add melody, flutes of various kinds were often used. Rattles and flutes are still quite popular in Mexico.

It is a disaster that the European conquests and missionary people destroyed their culture. Nevertheless we can hear their music. Their music has lived on in their hearts beyond the years. What we hear is the mixture of their traditional music and the modern one from Europe. The fusion appears not only in melody and rhythm, but also in musical instruments: rattles, drums, flutes, and conch-shell horns from indigenous people, and violins, guitars, harps, brass instruments, and woodwinds from Spain.

One good example is Mexican "mariachi", basis of many song styles, developed from folk music called "son" and influenced by Spanish elements. Trumpets were added for performance. Interestingly, Guatemala was one of the first regions to be exposed to European classical music.

Later, since African people were transported to this region, the music in Mesoamerica became more colorful. The "call and response" style of dialogue through singing is often heard (see Sect. 5.6). Marimbas were probably introduced from West Africa.

The reggae from Jamaica should not be forgotten. The reggae was originally a dance music. But it became a spiritual music for African-origin people related to the longing for Africa, ethnic identification and religious themes. In 2018 the reggae was added to the UNESCO's Representative List of the Intangible Cultural Heritage of Humanity.

Not only Jamaica but also many Caribbean islands created great mixtures of music, which have become popular all over the world. Famous examples are calypso and mérengue.

5.7.4 South America

In South America the Andean music is famous for its beauty and originality. It is played all along the backbone of this half continent formed by the mountains, valleys and planes of the Andes comprising parts of many states: Venezuela, Columbia, Ecuador, Peru, Bolivia, Chile, and Argentina. The Andean music is popular and has its core in rural areas and among indigenous populations. Original chants and melodies come from the areas of Peru and Bolivia inhabited by the Quechuas and Aymaras. Since 1960 it has a revival via the neo-folkloric "Nueva canción".

Andean instruments provide music with a special flavor: the zampona, the quena, the charango (all depicted in Figs. 6.6b, 7.1b and 6.3b) for melody and the bomgo (Fig. 6.2) for rhythm. Also the guitar should not be missing. The music is impressive, retains an individual charm and has been taken to the rest of the world. The group "Ruphay" has become known to have disseminated in 1968 Andean music to the stages on a global scale (Fig. 5.17).

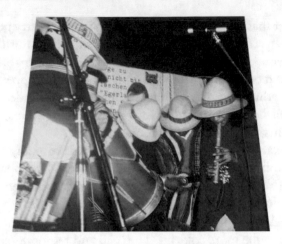

Fig. 5.17 A Bolivian music group, Ruphay plays Bolivian music with original indian instruments (CC BY-SA 4.0 by Rupprecht Weerth)

There have been famous voices to be heard around the globe singing songs of Andean origin with social and political messages: Victor Jara (from Chile) [17] and Mercedes Sosa (from Argentine) count among the most famous ones. Their songs have moved the minds and souls of many and still are.

In South America's east one encounters in Brazil another microcosmos of music genres, widely distributed and very rich in variations throughout this vast country [18]. Its evolution is closely connected with European colonization and the input of African slaves. Since the slaves had been deported from rather diverse regions of West Africa, their music traditions were not contained in a pure form but mixed such that a specifically Brazilian expression of African musical heritage developed. Music is strongly structured along polyrhythmicity, syncopation, percussion and improvisation together with call-and-response interplays.

The carnival in this country has a world-wide reputation. In its course Samba schools dance along their rhythms in all streets of the country. Beyond the omnipresent Samba rhythms one will hear the Bossa Nova and watch a multitude of other performances of popular dances. We find here a small afro-Brazilian universe in itself composed of rhythms, melodies, instruments, and dances, which to explore will not be feasible in the context of this chapter.

Fiesta Bahia - Samba

Cruisin Along - Bossa novaa

Areas far away from the center are the region of the Amazon forest and the Atlantic zone of Pará, where one dances the Carimbó, It is a most beautiful and very sensual dance. It involves only side to side movements and many spins and hip movements by the female dancer, who typically wears a rounded skirt. The music is mainly to the beat of Carimbó drums, approximately 1 m tall and 30 cm wide, made of a hollow trunk of wood and covered with a deerskin. The Carimbó style has formed the basis of some new rhythms like the Lambada.

Brazil has brought forth renowned composers (Antonio Carlos Gomes, Alexandre Levy, Heitor Villa-Lobos), guitarists (Turibio Santos, Baden Powell) and many other outstanding musicians.

We conclude our musical travel through Central and South America with a brief stop in Argentina with its beautiful glaciated mountains at the western Patagonian rim and the extended plains of the Gran Chaco and the Pampas. This country presented to the world a collection of dances full of character, rhythm and melody. The most popular one is the tango, which has spread internationally. It defines a cultural identity contributing to the self-conception of the people. Its lyrics are marked by nostalgia, sadness, and laments for lost love.

Tango dancing is essentially walking with a partner and the music in an improvised character without a basic step (see Fig. 5.18). It consists of a variety of styles that developed in different regions of the country.

Fig. 5.18 Tango by Astrid Nestvogel and Gregori Freiherr von Coburg-Maes (photo by Anna Arndt)

References

1. J. Diamond, *Guns, Germs, and Steel - The Fates of Human Societies* (W. W. Norton and Company, New York, 1999)
2. N.J. Conard, M. Malina, S. Münzel, New flutes document the earliest musical tradition in southwestern Germany. Nature **460**, 737 (2009)
3. W. Bühler, *Aristoxenos und Pythagoras - Ein elementarmathematischer Streifzug durch die Geschichte der musicalischen Slaken und Intervalle* (Peter Lang, Frankfurt a. M., 2017)
4. M. Schneider, Raga - Maqam - Nomos, in *Die Musik in Geschichte und Gegenwart*, Band 10 (Kassel, 1962)
5. http://boreale.konto.itv.se/smusic.htm, A brief introduction to traditional Sami song and the modern music
6. F. Hunter, The carnyx in Iron Age Europe. Ant. J. **81**, 77–108 (2001)
7. R. Yu-an, E.C. Carterette, W. Yu-Kui, A comparison of the musical scales of the ancient Chinese bronze bell ensemble and the modern bamboo flute. Percept. Psychophys. **41**, 547–562 (1987)
8. Y. Yung-Ching (ed.), *Chinese Folk Songs* (Arts Inc., New York, 1972)
9. F. Koizumi, *Studies on Japanese Traditional Music (written in Japanese)* (Ongaku no Tomosha, Tokyo, 2009)
10. T. Sakurai, Characteristics of Asian music and musical instruments. J. Acoust. Soc. Japan **54**, 651–656 (1998)

11. J. Lindsay, *Javanese Gamelan* (Oxford University Press, Oxford, 1979)
12. D. Christensen, Encyclopaedia Britannica (https://www.britannica.com/art/Oceanic-music)
13. B. Chatwin, *The Songlines* (Random House, New York, 2012)
14. K. Tsukata, *The Identity of African Music (written in Japanese)* (Ongaku no Tomosha, Tokyo, 2016)
15. F. Davis, *The History of the Blues* (Hyperion, New York, 1995)
16. R. Stevenson, *Music in Mexico: A Historical Survey* (Crowell, San Mariono, 1952)
17. J. Jara, *Victor: An Unfinished Song: Life and Music of Victor Jara* (Bloomsbury Publishing, London, 1998)
18. E. Morales, *The Latin Beat: The Rhythms and Roots of Latin Music from Bossa Nova to Salsa and Beyond* (Da Capo Press, Cambridge, 2003)

Part III
Physics of Musical Instruments

*The best is
still not good enough.*

—*modified from Steinways and Sons*

6

Musical Instruments

6.1 Systematical Classification of Musical Instruments

So many musical instruments have existed during the millenia and centuries, and so many exist still today all over the world. Lists have been compiled naming several hundreds of these and research is being performed to unravel relationships between them over time and distance.

In our approach we cannot focus on such an amazing plethora of musical utterances, but we will reduce our presentation on musical instruments in a quite pragmatic way, which starts with a general classification encompassing very many of the historic and actual cases used for music playing. Our first step into this matter is quite blunt.

In the classification given by Hornbostel and Sachs already in 1914, instruments from all continents have been considered [1]. Accordingly, we distinguish between the following four groups of instruments:

- Idiophones
- Membranophones
- Chordophones
- Aerophones

© Springer Nature Switzerland AG 2021
K. Tsuji et al., *Physics and Music*,
https://doi.org/10.1007/978-3-030-68676-5_6

6.1.1 The Idiophones

The sounding body itself serves for sound creation, because the material of the instrument provides the tone owing to its stiffness and elasticity, without needing strings or membranes. Jew's harp (see Fig. 1.5), bell, cymbal, xylophone, triangle, gong, chimes and others belong to this category.

A few characteristic examples are shown in the following pictures (Fig. 6.1).

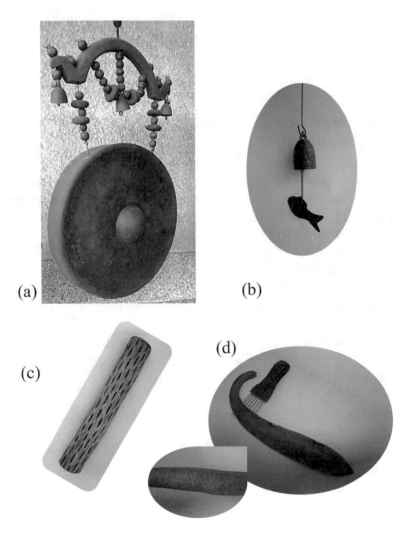

(a)   (b)   (c)   (d)

Fig. 6.1 a Gong from a Buddhist temple in Thailand, b bell in South East Asia, c rattle (shaker) made from a dry cactus stem with seeds inside (Chile), d scraped vessel with metal comb (Caribbean islands) (photos by KT)

6.1.2 The Membranophones

A vibrating membrane serves to produce sound together with a corpus for sound amplification (drum, tympani, bongo, mirliton).

Figure 6.2a shows bongos from Cuba. Another example is depicted in Fig. 6.2b: this is a kind of a snare drum with one membrane, as seen on the left side of this picture. Concert snare drums have several spring-like wires (see Fig. 8.18), while this drum has two lines of simple thread. However, even such a simple thread creates also the characteristic noise of snare drums.

(a)

(b)

Fig. 6.2 a Bongo set of single-skin conical drums, Cuba, late 19th century (CC BY-SA 3.0 by cralize), **b** drum from Morocco with threads behind the membrane (photo by KT, illustration by M. Covarrubias)

6.1.3 The Chordophones

Vibration of fixed chords leads to emission of sound; a corpus supports augmentation of sound (string instruments: violin, gamba, double bass, plucked: harp, guitar, lute, banjo, cembalo, keyboard: piano). See Figs. 6.3, 6.4, and 6.5.

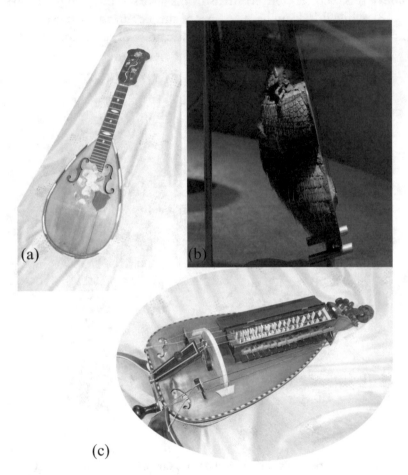

(a)

(b)

(c)

Fig. 6.3 a Mandoline (photo by KT in Volker Bley's gallery), **b** a traditional charango made of armadillo, Museu de la Música de Barcelona (CC BY-SA 3.0 by Enfo), **c** wheel fiddler (Hurdy-gurdy) (photo by KT in Volker Bley's gallery)

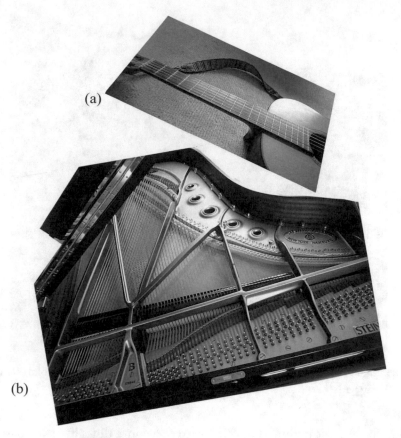

(a)

(b)

Fig. 6.4 **a** Guitar with frets for tone selection, **b** strings of a grand piano (photos by KT)

6.1.4 The Aerophones

Sound is produced by a vibrating column of air in a tube. Aerophones used in modern orchestras (wood wind instruments like flute, oboe, clarinet, saxophone, bassoon and brass instruments like trumpet, horn, trombone, tuba) and organ pipes are described in Chap. 7. Here we show in Fig. 6.6 some exotic instruments as examples.

Fig. 6.5 Members of the violin family. From left to right, 1/2 violin, 3/4 violin, 1/1 violin, viola (photo by KT)

A native American flute (Fig. 6.6a), also called love flute, has two chambers: one for collecting the breath of the player and a second one which creates sound. A block on the outside directs the breath from the first into the second chamber. The design of a sound hole at the proximal end of the sound chamber causes air from the player's breath to vibrate. A zampoña or siku (Fig. 6.6b, pan flute) is an instrument found all across the Andes. It consists of several bamboo tubes with different length, the end of which is closed. The gemshorn (Fig. 6.6c) is a member of the ocarina family mostly made from the horn of a chamois. (The name came from German "Gemse".) It dates back to at least the year 1450. The pointed end of the horn is left intact and a fipple plug

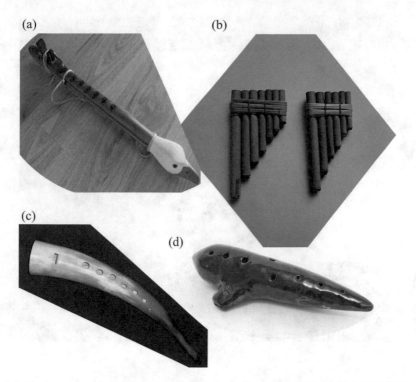

Fig. 6.6 Various aerophones. **a** native American flute (North America) (CC BY 2.5 by Jossi Fresco), **b** Peruvian zampoña (pan lute) (photo by KT) **c** gemshorn (CC BY-SA 3.0 by RiderOfRohan1981), **d** ocarina in Museu de la Música de Barcelona (CC BY-SA 3.0 by Patian)

is fitted into the wide end of the instrument. Ocarina (Fig. 6.6d), meaning "little goose", is a vessel flute with duct and finger holes of Mesoamerican origin. A mouthpiece projects from the body. Its spherical hollow body acts as a Helmholtz resonator (see Chap. 2). They have a distinctly an overtoneless sound.

Coffee Break

Serpent: Brass instrument with high intensity, in use from 1590 to the mid 19th century (illustration: KT, serpent: public domain)
    This special instrument belongs to the aerophones, but,..... Look at it! Isn't it too dangerous for a player?

6.2 Sound and Acoustic Waves

6.2.1 Excursion on Sound and Acoustic Waves

"Acoustics" is the branch of physics that deals with the study of all mechanical waves in gases, liquids, and solids including topics such as vibration, sound, ultrasound and infrasound. It is concerned with research and teaching of the origin, generation, spreading, influencing and analysis of sound, as well as effects, impacts and perception by our ears. In fact, hearing is one of the most crucial means of survival in the animal world, as speech is a distinctive characteristics of human development and culture. Accordingly, the science of acoustics spreads across many facets of human society—music, medicine, architecture, industrial production, and more.

Many sub-disciplines have been established in acoustics, such as physical acoustics, hydroacoustics, aeroacoustics, bioacoustics, electroacoustics, environmental noise, musical acoustics, psychoacoustics, speech, ultrasonics, room acoustics and audiometry.

For our purpose, some notions about the field of acoustics and physical quantities of importance are now being introduced for understanding the basic functioning of most musical instruments. So we give a brief overview on what we think is of relevance in this context.

6.2.2 History

The Chinese are known to have systematically treated tonal systems and tuning already in the 3rd millenium BC [2]. In ancient Greek civilization it was mainly Pythagoras (570–510 BC), who analyzed mathematically the relationship between the length of a monochord and its pitch (see Chap. 4). Chrysippos of Soli (281–208 BC) noticed the wave character of sound by comparison with the surface waves on water [3]. The Roman architect Vitruvius (ca. 80–10 BC) analyzed the sound propagation in amphitheaters [4].

After these early findings there follow many milestones, some of which are enumerated in short below:

- Leonardo da Vinci (1452–1519) detects that air is necessary as medium for sound propagation and that propagation occurs with finite speed.
- Vincenzo Galilei (1520–1591) describes the relationship between pitch and frequency [5].
- Isaac Newton (1643–1727) is the first to calculate the sound velocity on theoretical grounds.
- Leonard Euler (1707–1783) finds a wave equation for sound in its still valid form.
- Ernst Florens Friedrich Chladni (1756–1827) as founder of modern experimental acoustics presents his famous sound figures (Chladni figures) [6] (see Chap. 8).

In the 19th century an intensive activity in the field of acoustics started with Pierre-Simon Laplace (1749–1827), who worked on the adiabatic behavior of sound; Georg Simon Ohm (1789–1854), who postulated that our ear is able to separate sound into basic and harmonic components; Hermann von Helmholtz (1821–1894), who consolidated the field of physiological acoustics [7]; Lord Rayleigh (1842–1919) publishing his monumental work "Theory of Sound" [8].

The second half of this century brings along the construction of measuring and recording apparatus like the phonograph of Thomas Alvar Edison (1847–1931) or the absorption tube introduced by August Kundt (1839–1894).

The 20th century witnesses a broad application of now available theoretical knowledge to technical acoustics. Some keywords would be room acoustics, electro-acoustic transmission techniques, using ultrasound for locating objects under water (sonar), sound damping, medical applications via ultrasound, and many more. Sound recording and the telephone played important roles in a global transformation of society. Sound measurement and analysis reached new levels of accuracy and sophistication through the use of electronics and computing.

6.2.3 The Physics of Sound

Sound describes the propagation of smallest variations of pressure and density in an elastic medium (gases, liquids, solids). In solids sound can occur as transversal or longitudinal wave. In (ideal) liquids and gases sound propagates in form of a longitudinal wave, whereas transversal waves do not appear due to the lack of coupling forces. Particles can then exert forces only in longitudinal (spreading) direction.

Solids

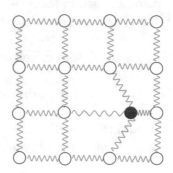

On neighbors of the displaced sphere forces can be imagined to act via springs.
Sound can occur as cross or longitudinal wave

Liquids and Gases

Sound propagates as a longitudinal wave.
Forces act between sound particles only in longitudinal direction

Sound particles are not molecules in the physical sense: they have neither defined physical or chemical properties nor the temperature-dependent kinetic behavior of ordinary molecules. They are very small compared to the wavelength of sound, so that their movement truly represents the movement of the medium in their locality.

6.2.4 Important Physical Quantities

A sound field is connected with spatial and temporal changes of the physical quantities—density ρ, pressure p and particle velocity $\mathbf{v}$, where ρ and p are scalars and $\mathbf{v}$ is a vector. Their total values (ρ_{tot}, p_{tot} and $\mathbf{v}_{tot}$) consist of constants for a state without sound field (ρ_0, p_0 and $\mathbf{v}_0$) and a variable part characterizing the effect of sound, according to

$$\rho_{tot} = \rho_0 + \rho$$
$$p_{tot} = p_0 + p$$
$$\mathbf{v}_{tot} = \mathbf{v}_0 + \mathbf{v}. \tag{6.1}$$

For describing the sound field it is sufficient to consider two quantities, conveniently $p(\mathbf{r}, t)$ and $\mathbf{v}(\mathbf{r}, t)$, where $\mathbf{r}$ and t are space vector and time.

6.2.4.1 Sound Pressure p

Sound pressure is an alternating pressure, superimposed on the static pressure (often air pressure) of the surrounding medium, p_0. The total pressure p_{tot} is a scalar with $p \ll p_0$ measured in Pascal (Pa), where 1 Pa = 1 Nm^{-2}, 1 bar = 10^5 Pa.

For a sinus tone (or pure tone),

$$p(t) = \hat{p} \cdot sin(2\pi f t) = \hat{p} \cdot sin(\omega t) \tag{6.2}$$

where $\hat{p}$ is sound amplitude, f is frequency (s^{-1}), and $\omega = 2\pi f$ is angular frequency (s^{-1}).

Distance dependency is

$$\tilde{p} = 1/r, \tag{6.3}$$

where $\tilde{p} = \hat{p}/\sqrt{2}$ is the effective value for a sinusoidal signal and r is the distance.

6.2.4.2 Sound Pressure Level (SPL) L_p

The SPL is defined by the logarithmic ratio of p^2 and p_0^2 in decibel (dB),

$$L_p = 10 \cdot log_{10}(p^2/p_0^2) = 20 \cdot log_{10}(p/p_0) \quad \text{(dB)}. \tag{6.4}$$

The reference value is set to

$$p_s = 20 \ (\mu Pa) = 2 \times 10^{-4} \ (Pa).$$

This is the auditory threshold at around 1 kHz.

As a rule of thumb a change by 10 dB is perceived as double or half of the sound intensity.

Table 6.1 Sound pressure and SPL of diverse sound sources

Source	Distance	Pressure in Pa	SPL in dB
Air jet	30 m	630	150
Rifle shot	1 m	200	140
Pain threshold	at ear	>100	>120
Running motor of air plane	30 m	90	120
Performing orchestra	30 m	11	100
Noisy street with much traffic	10 m	0.2–0.63	80–90
Passenger car	10 m	0.02–0.2	60–80
Brass orchestra	30 m	0.5	70
Persons speaking with each other	1 m	2×10^{-3}–6.3×10^{-2}	40–60
Calm room	at ear	2×10^{-4}–6.3×10^{-4}	20–30
Relaxed breathing, flattering of leaves	at ear	6.32×10^{-4}	10
Auditory threshold	at ear	2×10^{-4}	0

Sound pressures and SPLs of various sound sources are listed in Table 6.1.

The subsequent graphics (Fig. 6.7) shows curves of equal loudness as determined according to the sensitivity of our ear, which can detect sound waves with frequencies between 20 Hz and 20,000 Hz. Beyond the dB scale one has introduced the "phon". This is defined such that the dB scale for SPL and the phon scale for loudness level coincide at a sinus tone of 1000 Hz. The phon curves are valid only for tones of long duration.

As an example, we look at the curve for 80 phons: per definition, this value coincides at 1000 Hz with the sound pressure level 80 dB. On the other hand, at 100 Hz approximately 92 dB are required for the same loudness. Following the 60 phon loudness curve, a change of frequency from 1000 Hz to 100 Hz would cause an SLP change from 60 to 77 dB.

There are loudness levels even below the threshold of pain which for musical reception are no longer considered to be beautiful, but rather annoying with unexpected changes of tone color. This defines a threshold of discomfort, which should not be exceeded.

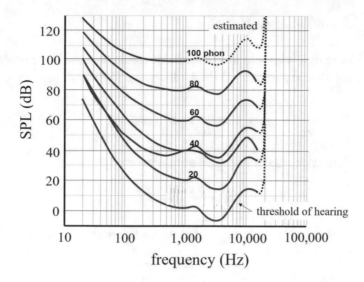

Fig. 6.7 Red: equal loudness curves for sound incident from the front according to ISO 226 (2003 revision), with threshold of hearing, blue: equal loudness curve at 40 phon from ISO standard before revision. The scale in the center shows loudness in phon at 1000 Hz. Threshold of pain for SPL $\geq$ 120 dB (public domain by Lindosland)

6.2.4.3   Particle Displacement ξ

We call ξ the momentary distance of a point P from its resting position (in general, a value x on the space coordinate). Thus, the sound mediated displacement is ξ, and its amplitude $\hat{\xi}$. As a space vector one would write the displacement as $\boldsymbol{\xi}$.

For a propagating sine wave one finds that $\hat{\xi}$ relates to the amplitude of the particle velocity (see below) according to

$$\hat{\xi} = \frac{\hat{v}}{\omega}, \tag{6.5}$$

with ω denoting the angular frequency.

6.2.4.4 Sound Particle Velocity v

This is the vector **v** with which the particles vibrate around their resting position. The momentary velocity is then, along one spatial coordinate x

$$\mathbf{v}_x = \frac{d\xi}{dt} = \dot{\xi} \quad (\text{ms}^{-1}) \tag{6.6}$$

A further derivative results in sound acceleration:

$$\mathbf{a} = \dot{\mathbf{v}} \quad (\text{ms}^{-2}). \tag{6.7}$$

Measurement is not easy, requires sensitive microphones or laser-Doppler-anemometry.

6.2.4.5 Sound Impedance Z

Z as the wave impedance (or resistance) of a medium is determined as the relation between sound pressure and particle velocity; in general a complex quantity due to phase shift.

But under far field conditions pressure and velocity are in phase, thus Z becomes real and

$$Z = p/v \quad (\text{Ns/m}^3) \tag{6.8}$$

with

$$v = |\mathbf{v}|.$$

Furthermore one has

$$Z = \frac{p}{v} = \frac{I}{v^2} = \frac{p^2}{I} = \rho \cdot c \tag{6.9}$$

where I is sound intensity, ρ is density and c is sound speed (see Table 6.2 below).

Table 6.2 Sound impedance of some media

Medium	Temperature (°C)	Z (Ns/m³)
Air	35	403.6
	20	413.6
	0	428.3
	−10	436.4
Water	10	1440.0
Iron		45600.0

6.2.4.6 Sound Energy W

$\widetilde{W}$ is the sum of kinetic and potential energy

$$\widetilde{W} = W_{pot} + W_{kin} = \int_V \frac{p^2}{2\rho_0 c^2} dV + \int_V \frac{\rho v^2}{2} dV \tag{6.10}$$

The sound energy density is the sound energy per volume unit, measured in (Ws/m³).

6.2.4.7 Sound Intensity I

I is an energy flux in sound fields - the sound power, which transverses an area per unit area. It is a directed quantity like the particle velocity.

$$I = p\mathbf{v} \quad (\text{W/m}^2) \tag{6.11}$$

Integrating over the considered area results in sound power (W).

6.2.5 Acoustic Wave

A major topic in acoustics is the understanding of wave propagation, because this forms the basis of communication between a sender (e.g. the voice) and a receiver (e.g. the ear) as shown in Fig. 1.4.

Acoustic waves (sound waves) are longitudinal waves that move by means of adiabatic compression and decompression of the medium in which they propagate. Longitudinal waves are waves that have the same direction of vibration as their direction of travel. Important quantities for describing acoustic waves are sound pressure p, particle velocity $\mathbf{v}$ and sound intensity I. They travel with the speed of sound c, which depends on the medium they are passing through.

They exhibit phenomena like diffraction, reflection and interference, but do not have any polarization, since they oscillate along the same direction as they move.

6.2.5.1 Planar Waves

The propagation of a sound wave in terms of sound pressure in one dimension is given by

$$\frac{\partial^2 p}{\partial \xi^2} - \frac{1}{c^2}\frac{\partial^2 p}{\partial t^2} = 0 \tag{6.12}$$

where p is sound pressure in Pa, ξ particle displacement in m, c speed of sound in m/s and t time in s.

The wave equation for particle velocity $\mathbf{v}$ has the same shape and is given (in one dimension) by

$$\frac{\partial^2 \mathbf{v}}{\partial \xi^2} - \frac{1}{c^2}\frac{\partial^2 \mathbf{v}}{\partial t^2} = 0 \tag{6.13}$$

For viscous media, more intricate models need to be applied in order to take into account frequency-dependent attenuation and phase velocity.

The general solution for the lossless wave equation is

$$p = R\cos(\omega t - kx) + (1 + R)\cos(\omega t + kx), \tag{6.14}$$

where ω is angular frequency in rad/s, k wave number in rad/m ($k = \omega/c = 2\pi/\lambda$, λ: wavelength) and R a coefficient without unit.

For $R = 1$ the wave becomes a traveling wave moving rightward, while for $R = 0$ the wave moves leftward.

An undamped standing wave can be obtained for $R = 0.5$, to be discussed in more detail below.

6.2.5.2 Standing Wave

A standing wave is a wave which oscillates in time but whose peak amplitude profile does not move in space. The peak amplitude of the wave oscillations at any point in space is constant with time, and the oscillations at different points throughout the wave are in phase. The locations at which the amplitude is minimum are called nodes, and the locations where the amplitude is maximum are called anti-nodes.

This phenomenon can arise in a stationary medium as a result of interference between two waves traveling in opposite directions. The most common cause of standing wave is the phenomenon of resonance, in which standing waves occur inside a resonator due to interference between waves reflected back and forth at the resonator's resonant frequency.

In one dimension, two waves with the same wavelength and amplitude, traveling in opposite directions will interfere and produce a standing wave. For example, a wave traveling to the right along a taut string held stationary at its right end will reflect back in the other direction along the string, and the two waves will superpose to produce a standing wave. This wave interaction forms the basis for the emission of tones in many musical instruments.

Line Theory

Let us consider wave propagation in one spatial direction x, as commonly found, e.g., for electromagnetic waves along lines or for acoustic waves in tubes (see previous paragraph).

For mathematical description we refer to Euler's formula

$$e^{ix} = cosx + isinx \quad (i : \text{ imaginary unit}) \tag{6.15}$$

from which follows:

$$cosx = \frac{1}{2}(e^{ix} + e^{-ix})$$
$$sinx = \frac{1}{2i}(e^{ix} - e^{-ix}) \tag{6.16}$$

This way, harmonic components can be conveniently expressed with the complex exponential function.

The sound field is composed of an incident wave

$$p_e = \hat{p}_e e^{i\omega t} e^{-ikx}$$
$$v_e = \frac{\hat{p}_e}{Z_s} e^{i\omega t} e^{-ikx} \tag{6.17}$$

and a recurrent wave

$$p_r = \hat{p}_e e^{i\omega t} e^{ikx}$$
$$v_r = -\frac{\hat{p}_r}{Z_s} e^{i\omega t} e^{ikx} \tag{6.18}$$

with ω: angular frequency, $k = \omega/c = 2\pi/\lambda$: wave number, and Z_s: specific impedance.

The sums for incident and reflected waves are:

$$p = (\hat{p}_e e^{-ikx} + \hat{p}_r e^{ikx})e^{i\omega t}$$

$$v = \frac{1}{Z_s}(\hat{p}_e e^{-ikx} - \hat{p}_r e^{ikx})e^{i\omega t} \qquad (6.19)$$

Choose a cross-section of the sound field with coordinate $x = 0$, where one has values $\hat{p}_0$ and $\hat{v}_0$. Then,

$$\hat{p}_0 = \hat{p}_e + \hat{p}_r \; ; \quad \hat{v}_0 = \frac{1}{Z_s}(\hat{p}_e - \hat{p}_r) \qquad (6.20)$$

or:

$$\hat{p}_e = \frac{1}{2}(\hat{p}_0 + Z_s \hat{v}_0) \; ; \quad \hat{p}_r = \frac{1}{2}(\hat{p}_0 - Z_s \hat{v}_0). \qquad (6.21)$$

This way we can use the hyperbolic functions

$$cosh(ikx) = \frac{1}{2}(e^{ikx} + e^{-ikx}) = cos(kx)$$

$$sinh(ikx) = \frac{1}{2}(e^{ikx} - e^{-ikx}) = isin(kx) \qquad (6.22)$$

to obtain

$$p = \left[\hat{p}_0 cosh(ikx) - Z_s \hat{v}_0 sinh(ikx)\right]e^{i\omega t},$$

$$v = \left[-\frac{\hat{p}_0}{Z_s}sinh(ikx) + \hat{v}_0 cosh(ikx)\right]e^{i\omega t}. \qquad (6.23)$$

Closed Tube

Suppose that the tube terminates with a reverberant (non-absorbent) closure at $x = 0$. At this point v must vanish ($\hat{v}_0 = 0$).

Pressure p_1 and particle velocity v_1 along the tube coordinate x would be described as

$$p_1 = \hat{p}_0 cosh(ikx)e^{i\omega t} = \hat{p}_0 cos(kx)e^{i\omega t}$$

$$v_1 = \frac{\hat{p}_0}{Z_s}sinh(ikx)e^{i\omega t} = \frac{\hat{p}_0}{Z_s}sin(kx)e^{i\omega t + \pi/2}. \qquad (6.24)$$

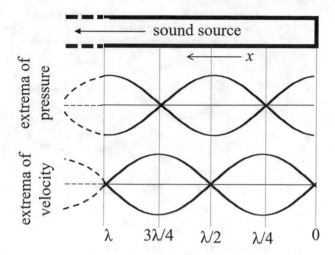

Fig. 6.8 Extrema of pressure and velocity in closed end tube

They represent a standing wave, where sound pressure and velocity vary harmonically all along the tube with a phase shift of $\pi/2$ rad.

Pressure is maximal at the end of the tube as shown in Fig. 6.8. It doubles because of interference of the incident with the reflective wave.

There are characteristic nodes and anti-nodes at distances of $x = n(\lambda/2)$ ($n = 0, 1, 2, \ldots$). The phase of maximal pressure values are shifted by $\lambda/4$ with respect to those of velocity.

Open Tube

If the tube is open at its end, sound pressure will break down there.

With $\hat{p}_0 = 0$, one derives

$$p_1 = Z_s \hat{v}_0 sinh(ikx)e^{i\omega t} = Z_s \hat{v}_0 sin(kx)e^{i(\omega t + \pi/2)},$$
$$v_1 = \hat{v}_0 cosh(ikx)e^{i\omega t} = \hat{v}_0 cos(kx)e^{i\omega t}. \qquad (6.25)$$

Here also a standing wave builds up. As compared to the previous case (Fig. 6.8) the sound and velocity fields are shifted by $\lambda/4$. A graphical presentation of its distribution along the length of the tube is drawn in Fig. 6.9.

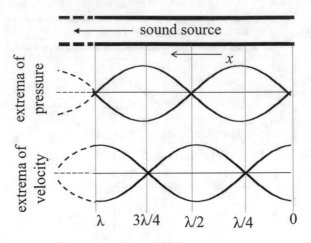

Fig. 6.9 Extrema of pressure and velocity in open end tube

6.2.5.3   Spherical Waves

We start with the relationships Eq. 6.16

$$cosx = \frac{1}{2}(e^{ix} + e^{-ix})$$

$$sinx = \frac{1}{2i}(e^{ix} - e^{-ix})$$

as derived from Euler's formula (see Eq. 6.15).

A sinusoidal function

$$p = \hat{p} \cdot sin(\omega t - kx + \varphi) \qquad (6.26)$$

can then be presented as:

$$p = Re\{\hat{p}e^{i\omega t}e^{\pm ikx}e^{i\varphi}\}, \qquad (6.27)$$

with

$$\hat{p} \rightarrow \hat{p}e^{i\varphi}.$$

Of great importance are spherical waves, when the sound field depends, after time, only on the distance r from a center. Being devoid of vortices, one can derive the particle velocity **v** from a potential Φ according to

$$\mathbf{v} = -grad\ \Phi, \qquad (6.28)$$

where,

$$grad = \left[\frac{\partial}{\partial x}e_x, \frac{\partial}{\partial y}e_y, \frac{\partial}{\partial z}e_z\right], \quad e_i : \text{unit vectors.}$$

From the time derivative there follows with the equation of motion

$$\rho\frac{\partial \mathbf{v}}{\partial t} = -grad\ p, \tag{6.29}$$

where

$$p = \rho\frac{\partial \Phi}{\partial t}.$$

The wave equation for Φ can then be obtained as

$$\Delta\Phi = \frac{1}{c^2}\frac{\partial^2\Phi}{\partial t^2}, \tag{6.30}$$

where Δ is the Laplace operator:

$$\Delta = \frac{\partial^2}{\partial x^2} + \frac{\partial^2}{\partial y^2} + \frac{\partial^2}{\partial z^2} \quad \text{with spatial coordinates } x, \ y, \ z.$$

In spherical coordinates this writes

$$\frac{\partial^2(r\Phi)}{\partial r^2} = \frac{1}{c^2}\frac{\partial^2(r\Phi)}{\partial t^2} \tag{6.31}$$

A particular solution is the spherical wave

$$\Phi = \frac{C}{r}e^{i\omega t}e^{-ikr}, \tag{6.32}$$

with C an amplitude constant.
From this one gets

$$p = \frac{i\omega\rho}{r}Ce^{i\omega t}e^{-ikr}$$

$$\mathbf{v} = \left(\frac{ik}{r} + \frac{1}{r^2}\right)Ce^{i\omega t}e^{-ikr}. \tag{6.33}$$

describing a radially propagating, divergent spherical wave.

6.2.5.4 Sound Velocity c

c is the propagation velocity of sound waves in (m/s) in a supportive medium; dependent on elasticity, density and temperature of the medium.

In liquids one has:

$$c_{liquid} = \sqrt{\frac{K}{\rho}} \tag{6.34}$$

where K is the compression modulus in (N/m^2).

In solids one has to consider the role of elasticity, Poisson number, density, and more.

For an ideal gas one may use the following formula

$$c_{idealgas} = \sqrt{\kappa \frac{p}{\rho}} = \sqrt{\kappa \frac{R_G T_{Abs}}{M}} \tag{6.35}$$

with R_G: universal gas constant (J/molK), M: molecular weight (kg/mol), T_{Abs}: absolute temperature (K), κ: adiabatic exponent.

For air, $\kappa = 1.402$, therefore, $c_{air} = 343$ m/s at $0\,°C$.

For real gases the adiabatic coefficient $\kappa = c_p/c_v$ may vary, especially due to temperature decrease or pressure increase.

With small temperature variation,

$$c_{gas} = (331.5 + 0.6 \cdot \vartheta) \quad (m/s) \quad : \vartheta = \text{temperature in }°C.$$

This approximation is valid to 99.8% in the temperature range from -20 to $40\,°C$.

Table 6.3 shows how the sound velocity of air depends on temperature (independent on p and ρ).

In Table 6.4 the sound velocity in various media is listed.

For wood instruments: see Table 6.5.

Table 6.3 Temperature dependence of sound velocity in air

Temperature ϑ (°C)	c_{air} (m/s)
35	352.2
25	346.4
20	343.5
10	337.6
0	331.0
−10	325.4
−20	319.1

Table 6.4 Sound velocity in various media

Medium	c (m/s) at 20 °C
Air	343
Helium	481
Hydrogen	1280
Oxygen	316[a]
CO_2	316
Water	1484
Ice (−4 °C)	3250
Glass	>4430
Al	6300
Au	3240
Cu	4060
Steel	5900
Fe	5170
Ti	6100
Diamond	18000

[a] at 0 °C

Table 6.5 Sound velocity in wood materials for musical instruments

Wood	c (m/s)
Alder	4400
Maple	4500
Ash-tree	4700
Limewood	5100
Pine tree	5500

6.3   From Sound Waves to Sound Frequencies

Many of the phenomena studied in science and engineering are periodic in nature. For example, as we explained above, the sound pressure in a closed tube, as well as in an open tube is periodic (see Figs. 6.8 and 6.9). These periodic functions can be written as a series of their constituents (fundamentals and harmonics) by a process called Fourier analysis, developed by the French mathematician Jean Baptiste Fourier (Fig. 6.10).

Fig. 6.10 French mathematician Jean Baptiste Joseph Fourier (1768–1830), engraved portrait by Louis-Léopold Boilly in the early 19th century (public domain)

6.3.1   Overview of Fourier Series

The Fourier series is an infinite series expansion involving trigonometric functions. A periodic wave function $f(t)$ of period T can be generally presented by an infinite Fourier series given by:

$$f(t) = a_0 + \sum_{n=1}^{\infty} a_n \cos\left(\frac{2\pi n}{T}t\right) + \sum_{n=1}^{\infty} b_n \sin\left(\frac{2\pi n}{T}t\right) \qquad (6.36)$$

where a_n and b_n are the Fourier coefficients, a_0 is the mean value.

Using the relationship for angular frequency

$$\omega = \frac{2\pi}{T} \tag{6.37}$$

Equation 6.36 can be written as,

$$f(t) = a_0 + \sum_{n=1}^{\infty} a_n \cos(n\omega t) + \sum_{n=1}^{\infty} b_n \sin(n\omega t) \tag{6.38}$$

Any periodic function can be expressed by Eq. 6.36 (or Eq. 6.38), by using appropriate Fourier coefficients a_n and b_n and the mean value a_0.

These coefficients are obtained by integration of $f(t)$ over period T:

$$\int_{-T/2}^{T/2} f(t)dt = \int_{-T/2}^{T/2} \left[a_0 + \sum_{n=1}^{\infty} a_n \cos(n\omega t) + \sum_{n=1}^{\infty} b_n \sin(n\omega t) \right] dt$$

$$= \int_{-T/2}^{T/2} a_0 dt + \sum_{n=1}^{\infty} \int_{-T/2}^{T/2} a_n \cos(n\omega t)dt + \sum_{n=1}^{\infty} \int_{-T/2}^{T/2} b_n \sin(n\omega t)dt \tag{6.39}$$

In general, for any integer m

$$\int_{-T/2}^{T/2} \cos(m\omega t)dt = \int_{-T/2}^{T/2} \sin(m\omega t)dt = 0. \tag{6.40}$$

Therefore,

$$a_0 = \frac{1}{T} \int_{-T/2}^{T/2} f(t)dt \tag{6.41}$$

In order to calculate a_n and b_n we multiply both sides of Eq. 6.39 with $\cos(k\omega t)$, where k is any integer, which is independent of n,

$$\int_{-T/2}^{T/2} f(t)\cos(k\omega t)dt$$

$$= \int_{-T/2}^{T/2} \left[a_0 + \sum_{n=1}^{\infty} a_n \cos(n\omega t) + \sum_{n=1}^{\infty} b_n \sin(n\omega t) \right] \cos(k\omega t)dt \tag{6.42}$$

and use the following relationships

$$\int_{-T/2}^{T/2} \cos(m_1\omega t)\cos(m_2\omega t)dt = \frac{T}{2}\delta_{m_1,m_2} \tag{6.43}$$

$$\int_{-T/2}^{T/2} \sin(m_1\omega t)\sin(m_2\omega t)dt = \frac{T}{2}\delta_{m_1,m_2} \tag{6.44}$$

$$\int_{-T/2}^{T/2} \cos(m_1\omega t)\sin(m_2\omega t)dt = 0, \tag{6.45}$$

where $\delta_{m_1,m_2} = 1$, only when $m_1 = m_2$, otherwise $\delta_{m_1,m_2} = 0$. The δ_{m_1,m_2} is known as Kronecker delta.

Since except the case $k = n$ all terms are 0, we obtain,

$$a_n = \frac{2}{T}\int_{-T/2}^{T/2} f(t)\cos(n\omega t)dt \tag{6.46}$$

$$b_n = \frac{2}{T}\int_{-T/2}^{T/2} f(t)\sin(n\omega t)dt, \tag{6.47}$$

where $n = 1, 2, 3, \ldots$.

6.3.1.1 Example 1: Saw-Tooth Signal

A saw-tooth signal with period 2π is described as a function $f(t)$ with

$$f(t) = t \quad (-\pi \le t < \pi) \tag{6.48}$$

and

$$f(t) = f(t + 2n\pi)$$

Inserting Eq. 6.48 and $T = 2\pi$ into Eqs. 6.41, 6.46 and 6.47, we obtain

$$a_0 = \frac{1}{2\pi} \int_{-\pi}^{\pi} t \, dt = 0$$

$$a_n = \frac{2}{\pi} \int_{-\pi}^{\pi} t \cdot \cos(nt) dt = 0$$

$$b_n = \frac{2}{\pi} \int_{-\pi}^{\pi} t \cdot \sin(nt) dt = \frac{2}{\pi} \left[t \frac{-\cos(nt)}{n} \right]_0^{\pi} - \frac{2}{\pi} \frac{1}{n} \int_0^{\pi} -\cos(nt) dt$$

$$= \frac{2}{\pi} \left(\pi \frac{-\cos(n\pi)}{n} - 0 \right) - \frac{2}{n\pi} \left[t \frac{-\sin(nt)}{n} \right]_0^{\pi} = 2 \frac{(-1)^{n-1}}{n}$$

Therefore, Eq. 6.36 becomes

$$f(t) \approx \sum_{n=1}^{\infty} 2 \frac{(-1)^{n-1}}{n} \sin(nt) \tag{6.49}$$

Now we will show how the Eq. 6.49 approximates the saw-tooth signal with increasing the number of n. It can be rewritten as

$$f(t) = 2\sin t - \sin 2t + \frac{2}{3}\sin 3t - \frac{1}{2}\sin 4t + \frac{2}{5}\sin 5t - \cdots . \tag{6.50}$$

The first approximation (taking only the first term) is a sine curve (the red curve of Fig. 6.11b), which does not yet look like the saw-tooth signal. However, when the first 5 terms are used, the curve approximates the saw-tooth signal already quite well (the red curve of Fig. 6.11c). If we take an infinite number of terms, the graph will look like a set of saw teeth (Fig. 6.11c). We say that the infinite Fourier series converges to the saw tooth curve (Fig. 6.11a).

There is an interesting connection between the saw-tooth signal and the bowing of a violin string, because the stick-slip situation during bowing results in a kind of saw-tooth shaped vibration of the string [7] (see Sect. 8.1.6).

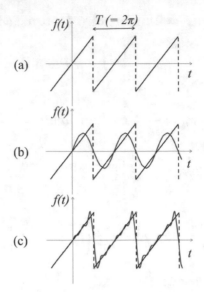

Fig. 6.11 a A saw-tooth signal, **b** red curve presents the 1st approximation, **c** red curve presents the 5th approximation

6.3.1.2   Example 2: Square Wave

A square wave with period $2(T = 2)$ can be described as a function $f(t)$:

$$f(t) = 0, \text{ if } -1 \le t < 0,$$
$$f(t) = 1, \text{ if } 0 \le t < 1, \tag{6.51}$$
$$f(t) = f(t + 2n).$$

From Eqs. 6.41, 6.46 and 6.47

$$a_0 = \frac{1}{2} \int_{-1}^{1} f(x) \cdot 1 dt = \frac{1}{2} \int_{0}^{1} 1 \cdot 1 dt = \frac{1}{2} [t]_0^1 = \frac{1}{2} \tag{6.52}$$

$$a_n = \frac{2}{2} \int_{-1}^{1} f(t) \cos(n\pi t) dt = \int_{0}^{1} 1 \cdot \cos(n\pi t) dt$$
$$= \left[\frac{\sin(n\pi t)}{n\pi} \right]_0^1 = 0 - 0 = 0 \tag{6.53}$$

$$b_n = \frac{2}{2}\int_{-1}^{1} f(t)sin(n\pi t)dt = \int_{0}^{1} 1 \cdot sin(n\pi t)dt$$

$$= \left[\frac{-cos(n\pi t)}{n\pi}\right]_{0}^{1} = \frac{(-cos(n\pi) - (-1))}{n\pi} = \frac{1 - (-1)^n}{n\pi} \quad (6.54)$$

Therefore, Eq. 6.36 becomes

$$f(t) \approx \frac{1}{2} + \sum_{n=1}^{\infty} \frac{1 - (-1)^n}{n\pi} sin(n\pi t) \quad (6.55)$$

Now we will show how Eq. 6.55 approximates the square wave with increasing the value of n. This equation can be rewritten as

$$f(t) = \frac{1}{2} + \frac{2}{\pi}sin(\pi t) + \frac{2}{3\pi}sin(3\pi t) + \frac{2}{5\pi}sin(5\pi t) + \frac{2}{7\pi}sin(7\pi t) + \cdots \quad (6.56)$$

The first approximation (taking only the first two terms) is a sine curve (the red curve of Fig. 6.12b), which does not yet look like the square wave. However, when the first 5 terms are used, the curve approximates the square wave to

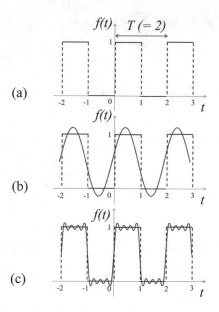

Fig. 6.12 a A square wave, **b** red curve presents the 1st approximation, **c** red curve presents the 5th approximation

some extent (the red curve of Fig. 6.12c). If we take an infinite number of terms, the graph will look exactly like a square wave (Fig. 6.12a).

6.3.2 Transformation from Sound Waves to Sound Spectra

6.3.2.1 Harmonics and Line Spectrum

Using the following trigonometric relationship

$$a \cdot cos\theta + b \cdot sin\theta = R \cdot cos(\theta - \alpha),$$

where

$$R = \sqrt{a^2 + b^2} \ \ and \ \ \alpha = \arctan\frac{a}{b},$$

the Fourier series (Eq. 6.38) can be written in harmonic form

$$f(t) = a_0 + \sum_{n=1}^{\infty} [a_n cos(n\omega t) + b_n sin(n\omega t)]$$

$$= C_0 + \sum_{n=1}^{\infty} C_n cos(n\omega t - \varphi_n), \tag{6.57}$$

where

$$C_0 = a_0, \quad C_n = \sqrt{(a_n)^2 + (b_n)^2}, \quad \varphi_n = \arctan\frac{a_n}{b_n}.$$

Equation 6.57 can be rearranged to

$$f(t) = C_0 + C_1 cos(\omega t - \varphi_1) + C_2 cos(2\omega t - \varphi_2) + C_3 cos(3\omega t - \varphi_3) + \cdots.$$
$$\tag{6.58}$$

Here,

$C_1 cos(\omega t - \varphi_1)$ is the **fundamental** or the **first harmonic**
$C_2 cos(2\omega t - \varphi_2)$ is the **second harmonic**
$C_3 cos(3\omega t - \varphi_3)$ is the **third harmonic**
....
$C_n cos(n\omega t - \varphi_n)$ is the **nth harmonic**

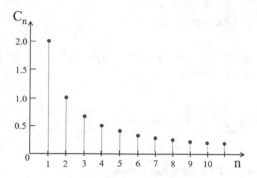

Fig. 6.13 Line spectrum for saw-tooth function

The plot showing each of the harmonic amplitudes for the saw-tooth signal is called its line spectrum (Fig. 6.13). Note that waves with discontinuities such as the saw tooth and square wave have spectra with slowly decreasing amplitudes, since their series have strong high harmonics. Their 10th harmonics and higher ones will often have amplitudes of significant value compared to the fundamental.

The high harmonics thus appear, when the temporal signal displays "steep" slopes, which cause high frequencies to play an important role in the spectrum. This is the case, for instance in the excitation of a violin cord by the bow.

Playing harmonics on a guitar: If you just lightly touch a string with the left hand and then pluck it, you hear a high pitched sound called a harmonic (see Fig. 6.14). For the violin this corresponds to playing "flageolett". In wind instruments overblowing creates harmonics. Music sounds "in tune", because the harmonics contained in each note sound "right" in relation to certain other notes.

Fig. 6.14 Playing harmonics on a guitar (photo by KT)

6.3.2.2   Another Example

The following Fourier series with $\omega = 1$ (period $T = 2\pi$) is shown in the plot of Fig. 6.15a. Its function $f(t)$ is:

$$f(t) = \frac{2}{\pi} + \sum_{n=1}^{\infty} \frac{1}{2n-1}\cos(nt) + \sum_{n=1}^{\infty} \frac{(-1)^n}{2n}\sin(nt) \qquad (6.59)$$

It can be rewritten in the harmonic form:

$$f(t) = \frac{2}{\pi} + \sum_{n=1}^{\infty} C_n \cos(nt - \varphi_n), \qquad (6.60)$$

where

$$a_n = \frac{1}{2n-1}, \qquad b_n = \frac{(-1)^n}{2n}$$

and

$$C_n = \sqrt{(a_n)^2 + (b_n)^2}$$

The resulting line spectrum is shown in Fig. 6.15b.

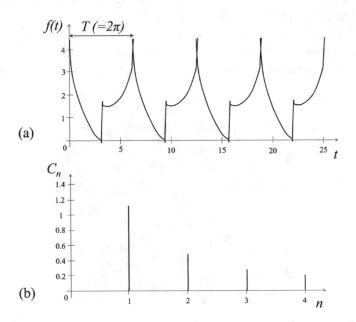

Fig. 6.15 a Wave form of the Fourier series presented in Eq. 6.59, b its line spectrum

6.3.2.3 Fourier Transform

Above we showed by using Fourier series how a periodical time function can be described as a function of frequency. More generally, there is the Fourier (integral) transform which decomposes a function of time (time domain) into its constituent frequencies (frequency domain), thus opening a much more detailed frequency analysis of complex signals.

The Fourier transform of a function f, denoted $\hat{f}$ is

$$\hat{f}(\xi) = \int_{\infty}^{\infty} f(x)e^{-2\pi ix\xi} dx \tag{6.61}$$

for any real number ξ.

When the independent variable x represents time, the transform variable ξ represents frequency (e.g. if time is measured in seconds, then frequency is in Hz). This is the inverse Fourier transform:

$$f(x) = \int_{\infty}^{\infty} \hat{f}(\xi)e^{2\pi ix\xi} d\xi \tag{6.62}$$

for any real number x.

Using the Fourier transform or inverse Fourier transform one can obtain a sound spectrum from any sound wave or vice versa.

6.3.2.4 Application to Musical Sound

As described in Chap. 4, we hear a specific color (timbre) when an instrument is played or a human sings. The harmonic series is characteristic for each sound. Figure 6.16 shows the wave forms of flute and violin, and their corresponding harmonic series obtained by using Eq. 6.61.

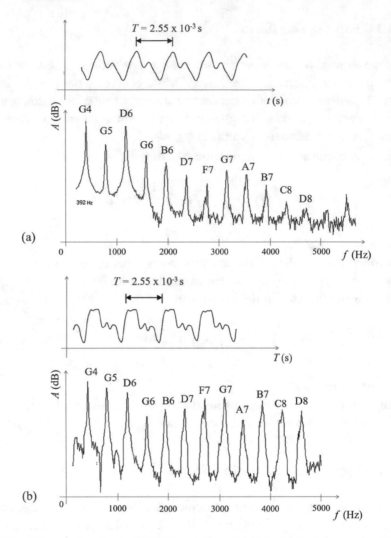

Fig. 6.16 a Wave form of sound and its spectrum of flute, **b** wave form of sound and its spectrum of violin for G4 (392 Hz), the period $T = 2.55 \times 10^{-3}$ s (data source of the line spectra: University of New South Wales Physics Department [9])

References

1. E.M. von Hornbostel, C. Sachs, Systematik der Musikinstrumente: Ein Versuch. Z. Ethnol. **46**, 553–590 (1914)
2. J. Jie, *Chinese Music* (Cambridge University Press, Cambridge, 2011)
3. T.L. Heath, *A History of Greek Mathematics, Vol 1: From Thales to Euclid* (Clarendon Press, Oxford, 1921)

4. Vitruvius *Ten Books on Architecture*, ed. by I. Rowland (Cambridge University Press, Cambridge, 1999)
5. V. Galilei, *Fronimo Dialogo di Vincentio Galilei Nobile Fiorentino, Sopra L'arte del Bene Intavolare, et Rettamente Sonare la Musica* (G. Scotto, Venice, 1584)
6. E.F.F. Cladni, *Entdeckungen über die Theorie des Klanges* (Discovery on the Sound Theory) (Weitmanns Erben und Reich, Leipzig, 1787)
7. H. Helmholtz, *Die Lehre von des Tonempfeindungen, als physiologische Grundlage für die Theorie der Musik* (Friedrich Vieweg und Sohn, Braunschweig, 1863)
8. J.W.S. Rayleigh, *The Theory of Sound*, vol. 1 (Macmillan and Co., London, 1877)
9. https://newt.phys.unsw.edu.au/music/

7

Selected Instruments 1—Aerophones

7.1 Classification of Aerophones

In the rich variety of musical instruments developed in cultures all over the globe one finds some creations that have gained special popularity and are played today more frequently than many others. From these we will now select a few for more detailed study. In this chapter aerophone families are described.

Since there are many aerophones with different physics/mechanics, we classify them briefly before describing them individually.

- Woodwind instruments
  - without reed
    - with windway: recorder (open end), whistle (closed end)
    - without windway: transverse flute (open end), ocarina (closed end)
  - single-reed instruments: clarinet, saxophone
  - double-reed instruments: oboe, English horn (cor anglais), bassoon
- Brass instruments: horn, trumpet, trombone, tuba
- Organ pipe

7.2 The Flute

Among the aerophones, the flute is one of the most ancient instruments used in many cultures including Paleolithic tribes (see Chap. 1 Fig. 1.2). During the millenia different prototypes of flutes have been developed and used. Some

© Springer Nature Switzerland AG 2021
K. Tsuji et al., *Physics and Music*,
https://doi.org/10.1007/978-3-030-68676-5_7

of them are shown in Fig. 6.6 of Chap. 6. A predecessor was in some sense the "simple" whistle, sound producer relevant for the flute, which has been employed since the stone age as a signaling instrument, for instance on hunting trips.

7.2.1 Windway

A flow of air across an opening produces sound, which is the basic principle of a flute. The player is required to blow air across a sharp edge. There the air stream is split and then acts upon the air column contained within the flute's hollow causing it to vibrate and produce sound. Some flutes possess a "windway", which is a narrow air channel or duct directing and splitting the air stream at the sharp edge of the labium lip. With such a windway sound is created very easily just by blowing gently air into the opening.

Other flutes have no windway. To create sound without windway requires more sophisticated techniques. Usually a part of the air opening is cut, forming a sharp edge. A strong air stream should be focused on this sharp edge by using the lips or the tongue, so that the air is split.

Figure 7.1 shows a recorder (flute with a windway) and an Inca flute (without a windway).

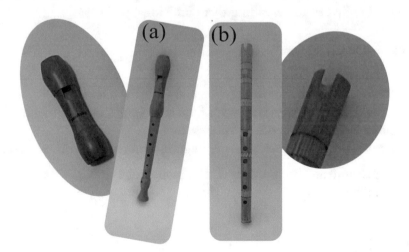

Fig. 7.1 Left: Recorder and its head with windway, right: Quena (Inca flute) and its opening with a notch at the upper rim of the tube (photos by KT)

Furthermore, there are two sub-families of the flute family: "open" flutes (both ends are open) and "closed" flutes (only one end is open). This implies different boundary conditions with consequences for the types of acoustic waves emanating from the instrument. Some basic properties have been described in Chap. 6. If the length of the tube is identical, the created sound is lower for closed flutes than for open flutes.

7.2.2 The Recorder

Many children have their first encounter with practising music when they learn to play the recorder. They may not know that this a descendant of the prehistoric flutes made from animal bones and that playing the flute was already a topic in the Bible. Wooden flutes are documented for the bronze age (1000 BC). In any case, the modern flute version called recorder is today in common use all over Europe.

Since the Renaissance one has used a family of recorders with various lengths:
Flute in F5 (sopranino)—20 cm
Flute in C5 (soprano)—32.5 cm
Flute in F4 (alto)—47 cm
Flute in C4 (tenor)—62 cm
Flute in F3 (bass)—96 cm.
(Here, "Flute in F5" means the lowest tone of this flute is F5.)

Construction of these flutes of different length was asked for by Michael Prætorius (1571–1621) [1]. Especially the alto recorder was appreciated for soloistic playing and for chamber music. With an ensemble of all family members the entire spectrum of Baroque to modern compositions can be played.

Commonly, there is a difference between the sound of a tone on a recorder and its notation, e.g., if sopranino and soprano recorder are tuned in the G clef, the tones sound one octave higher. In fact, the parameters for the recorder correspond to one octave deeper sounding intervals of the other music instruments. The soprano recorder would thus be comparable to the piccolo flute. A possible explanation of this "shift" in the recorder names may lie in a relative lack of harmonics in the flute's timbre, which makes recorder tones appear less bright than the equally high tone played by the violin.

7.2.2.1 Dynamics of the Air Stream and Sound Creation

A recorder consists of three sections: head, middle section (with internal bore and finger holes) and foot. The fipple plug in the mouthpiece is almost always

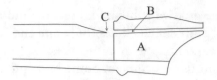

Fig. 7.2 Sections of a flute. **A**: wooden fipple plug, **B**: duct, **C**: bladed edge

made from cedar wood, the corpus from hard wood—without much influence on the sound.

Head of the recorder is where the sound is produced. It is, in principle, just a whistle and also similar to the function of an organ pipe (see Sect. 7.6). As shown in Fig. 7.2, there is a fipple plug (A) with a "ducted flue" windway above it in the mouthpiece of the flute, which compresses the player's breath, so that it travels along the duct (B). Being directed against a hard, bladed edge, the labium lip (C), a break of the laminar flow leaving the duct is produced. This results in a mixture of air stream, vortices and pressure waves, from which complex vibrations and a melodious sound arise.

We have observed by interferometric techniques [2] the air flow following a blow along the duct of a head of flute.

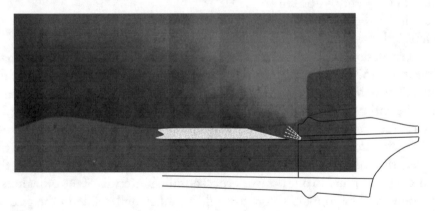

Fig. 7.3 Air coming out from the head of a flute observed with a Mach-Zehnder interferometer. Experiments were carried out by H. Kleine and K. Tsuji in the shock wave laboratory, RWTH Aachen University

Flux in viscous fluids including gases becomes easily unstable, that is, it does not maintain its initial direction. A blow of air exiting through a thin channel will diverge and tends to become turbulent forming a kind of vortex street as seen in Fig. 7.3. Putting an obstacle with a sharp edge into the air stream, close enough to the wind way where the stream is still about laminar (and not yet turbulent), the stream begins to oscillate.

This forms the basic coupling to the tube resonator leading to a resonance of the air column inside the instrument.

More precisely, the air flow formed in the duct starts to vibrate at the labium lip such that the air flows alternatively to the inner part of the flute and outward. The frequency of this motion is determined by the air column in the middle and lower parts of the tube.

This produces a rough edge tone. For a duct width of $2b$, without considering compressibility or viscosity, Lord Rayleigh [3] found a relationship between the streaming velocity V of the moving jet and the propagation velocity u of a small disturbance with wavelength λ according to

$$u = V/(1 + coth(kb)), \tag{7.1}$$

where $k = 2\pi/\lambda$ depends on the distance between duct and edge. The role of the jet dynamics may be of relevance on this short distance. (A further discussion of Eq. 7.1 see the book of Fletcher and Rossing mentioned in the Preface.) So that one can regulate the pitch of the edge tone by changing the blowing pressure.

7.2.2.2 Generator/Resonator

The longer middle bore with finger holes serves as a resonator for longitudinal waves of air pressure along the central axis of the tube.

The excitatory air stream is guided across an opening in the resonator to the rim of the labium such that the stream is partially led to the inner part of the resonator (the long tube). This way the pressure waves are induced in the resonator with the frequency of the generator. The conditions at the boundaries of the resonator are decisive about which pressure-generated sound emerges.

We consider a tube with open ends. As already described in Chap. 6, standing waves evolve with constant sound pressure at the ends distributed according to different modes. A few examples are depicted in Fig. 7.4.

These are standing waves caused by reflection at the open lower end of the tube (in Sect. 6.2.5). Not the air moves in the tube, but the pressure variations do. The blown air serves only for the excitation. Consequently there is no overall net motion of air along the inside of the bore of the instrument.

For a stable tone to form the phase difference $\Delta\varphi$ between oscillations of the sound particle velocity and the sound pressure inside the tube is $\pi/2$. Thus,

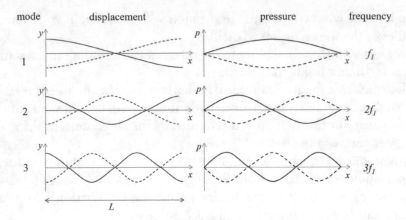

Fig. 7.4 Amplitude of displacement y and pressure p at three modes in a tube with open ends. Pressure at tube ends is constant

the system reaches resonance, energy is fed into the resonator, the boundary conditions for standing waves in a tube with open ends are fulfilled.

7.2.2.3   End Correction of Resonator Length

Above we learned that the fundamental frequency f_1 of the resonance body of length L is

$$f_1 = \frac{1}{2L}.$$

However, what we hear from the flute of length L is somewhat deeper, indicating the actual frequency is slightly larger than f_1. The reason is that the generated standing wave inside the bore also propagates to some extent as a spherical wave into the outer space, radiating part of its energy, as if the reflection of the resonator wave takes place at the location ΔL away from the end of the tube.

Therefore, for constructing a flute or other wind instruments, the end correction is very important. One has to add ΔL (the so-called outlet correction) to the theoretical value of L.

Rayleigh calculated the value ΔL for various cases [3] in 1877. Much later, according to a further calculation, Levine/Schwinger [4] gave a realistic value of

$$\Delta L = 0.61R,$$

where R is tube radius.

7.2.2.4   Finger Holes

Recorders have, as a rule, 7 finger holes on the upper side of the tube and one on its back side (for overblowing). The basic tone is played when all holes are closed. When reaching the 3rd tone of the 2nd octave and the back side hole is only half covered, the tone switches to the octave of the basic tone. By opening of the finger holes one shortens the vibrations in the tube and the last open hole is then deciding on the density distribution in the tube. This way one can produce higher harmonics which determine the timbre of the tone. In fact, for a recorder the harmonics contribution is not very pronounced, so the tones are quite "basic". Producing harmonics depends on factors such as tone generation, blow pressure, form of the bore (cylindrical or conical), material (small influence), or irregularities on the wall of the bore.

An interferometric picture of a flute's resonator tube is shown in Fig. 7.5. When all holes are open, air exits all of them and also the open end of the flute. The fluxes are laminar, if the tone is not overblown. This air carries the tone outward, so it can be heard.

Fig. 7.5 When all holes are open, air comes out from all holes observed with a Mach-Zehnder interferometer. Air exiting at the end is laminar when it is not overblown. Experiments were carried out by H. Kleine and K. Tsuji in the shock wave laboratory, RWTH Aachen University

7.2.3   The Transverse Flute

Among many kinds of flutes an outstanding and very successful example is the transverse flute, which nowadays is a permanent member of symphony orchestras. It is a side-blown woodwind instrument with a long history [5]: A transverse held to the left was depicted on an Etruscan relief of about 200–100 BC. Illustrations from the 11th century show flutes played to the right.

There is a widespread use of such flutes by indigenous people, for instance when religious cult is practised.

The transverse flute arrived in Europe, where the recorder had been most common, from Asia via the Byzantine Empire in the 12th/13th centuries. It became used in court music, initially only in France and Germany. Renaissance offered a more general appreciation of flutes in chamber music often playing the tenor voice in musical ensembles. They were available in several sizes, mostly made from one section with a cylindrical bore producing a relatively soft sound.

During the Baroque period, the transverse flute was redesigned (now often called traverso). Then it was made in three sections with a conical bore. Due to its wider range and more penetrating sound it entered into orchestral music. However, with the romantic era, flutes again lost favor and symphony orchestras rather featured brass and strings.

With developments initiated by Theobald Böhm, who in 1847 patented his system, dimensions and key system of the modern transverse flute were established [6].

We are dealing here with a flute, which is an open cylinder with a side-blown "embouchure hole" over which a rapid jet of air is blown. It is an instrument without a windway. Its length: 67.5 cm, its outer diameter: 1.7 cm. The playing range is from B3 (or C4) to F7.

The modern transverse flutes are mostly made of metal (instead of the previously used wood). However, the material appears to play a subordinate role in relation to tonal impressions. Special preferences for gold or platinum rest predominantly on psychological effect, which, of course, may influence the performance technique of the flautist. Such a modern flute and a smaller transverse flute "piccolo" are shown in Fig. 7.6.

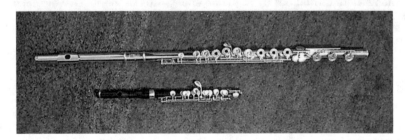

Fig. 7.6 Top: a modern transverse flute, made of metal, bottom: a piccolo made of wood (photo by KT in B. Geiger's house)

A piccolo has almost the same fingerings as its larger sibling, the standard transverse flute: length 32 cm, cross section 1 cm. Its tuning is one octave higher. So it covers the range from C5 to D8.

In early times the piccolo appeared in military music together with small drums. Later it was developed according to principles of the transverse flute, parallel to the Böhm-system. In orchestras a piccolo as "flute III" adds sparkle and brilliance to the overall sound. Supporting the very bright tone of the piccolo, one notices harmonic components up 10,000 Hz which nevertheless do not give metallic sharpness to the tone, since the frequency separation of the partials is very large due to the high pitch.

Although once made of wood, glass or ivory, piccolos are made today from plastic, resin, brass, silver, and a variety of hardwoods (e.g., granadilla from tropical regions in Central and South America).

7.2.3.1 Structure and Function

The flutes consist of three parts (similar to the recorder): head joint, body, and foot joint.

Head Joint

The upper third of the tube has the crown at its end helping to keep the head joint cork positioned at the proper depth, as shown in Fig. 7.7, left. The main function is assigned to the embouchure hole surrounded by the lip plate where the player initiates the sound. This plate contacts the player's lower lip, allowing positioning and direction of the air stream. In a cross section we can see the position of both lips and a sketch of flow lines of the air blown across the hole, being deflected alternately up and down, out of and into the bore of the flute (Fig. 7.7 right). This flow pattern forms the basis of sound creation.

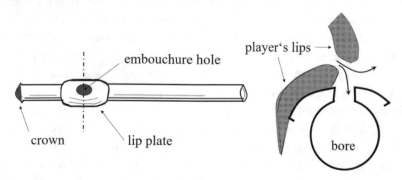

Fig. 7.7 Left: head of a transverse flute, right: cross section of the embouchure with lip position and air jet being deflected alternately up and down

The head joint is, as opposed to the rest of the tube, not cylindrical, but conical, which has a pronounced influence on the timbre.

Body

This is the middle section of the flute with the majority of keys. Along the body of the tube there are typically 16 circular tone holes. The pitch of the flute is changed by opening or closing keys that cover these tone holes. One may use closed-hole keys or open-hole keys with a perforated center, which are preferable for precise playing.

Foot Joint

The last section of the flute may be constructed as to determine the lowest playable note. The C-foot enables the player to produce the middle C (C4), the B-foot goes down to B3, and due to increased length of the resonating tube the sound appears warmer and more intense.

7.2.3.2   Sound Spectra

The tone of transverse flutes is characterized by the very uniform overtone structure of the spectra. The fundamental is the most pronounced of all partials for the entire range of the instrument (see Fig. 6.16a). For no other orchestral instrument is this characteristics as clearly marked. Above the fundamental the intensity of the overtones drops quite linearly with increasing frequency. The intensity relationship between the lower partials, and thus the tone color, can be varied by the performer through the blowing pressure, the degree of coverage of the mouth hole, and the blowing direction. Raising the blowing pressure leads to a better coincidence of the first overtones relative to the fundamental. The tones thus brighten in color. Reduced coverage of the mouth hole leads to a softening of the air stream at its edge. Finally, the blowing direction determines the strength relation between even and odd numbered partials.

The flute sound is perceived as having particular tonal beauty, when the fundamental and octave are about equal in strength, and the twelfth is weaker by about 10–15 dB. As to noise, it often appears in overblown notes and may result from statistical excitation of "unused" resonances.

The dynamics of the flute can be influenced in relatively narrow limits (about 6 dB) without changing the tone color by variation of the lip opening. The playable dynamic range depends especially strongly on pitch. The low register with a *pp* around 67 dB and a *ff* in the neighborhood of 86 dB is quite weak, in the highest octave the sound power level can reach 100 dB for *ff*, while for *pp* it can hardly be lowered below 83 dB [7].

The dynamic range of the piccolo is particularly narrow, covering just 15 dB. At *pp* it is therefore quite loud in comparison to other instruments. But an *ff* produces values between 93 and 103 dB, so the piccolo may stand out in a full instrumentation orchestra sound. Just listen to symphonies of Dmitri Shostakovich.

The flute requires the longest time for tonal development. There are so-called preliminary tones, which are formed, upon initial blowing, by higher resonances. These have a duration of about 50 ms and their intensity is perhaps 10 dB higher than the subsequent stationary state. The strength of the preliminary tones depends on the sharpness of the attack, but they have by no means a negative characteristics. It enhances the perception of the attack, important in view of the relatively long initial transients for flute staccatos. This participates in the tone effect known as "flutter tongue", as used, for example, by R. Strauss in a tone painting of his Don Quixote. Here the individual tongue beats of the "flutter tongue" follow each other with a frequency 25 Hz and with an amplitude modulation of 15 dB.

The decay process for the flute can be influenced to some extent by the performer, which is unique among wind instruments. The decay time (above 60 dB) is about 125 ms. A soft termination of the tone prolongs this time to about 200 ms.

7.2.3.3 Physical Aspects

Finally, when discussing the physics of sound creation in a transverse flute, we rely on the principles already pointed out in our presentation about the recorder and tube resonances. Blowing into the embouchure hole causes a split stream of air which is the source of air vibrations in the body of the flute and creates standing waves there.

The transverse flute is open at both ends: at the far end and also at the embouchure hole, where the player leaves a part of the hole open to the atmosphere (see Fig. 7.8a). We know from our discussion in Chap. 6 that with the total pressure at the open ends being approximately atmospheric pressure, pressure nodes form there and a standing wave may develop between these nodes with a maximum variation right at the middle. The wavelength of this standing wave is just about twice the length of the flute.

The deepest note played on the Böhm flute is B3 or C4. This would sound, if all finger holes are closed. By blowing harder one can produce a series of higher notes corresponding to standing waves with shorter wavelengths. In fact, the harmonic series of overtones can be excited this way, up to the 10th of these resonances.

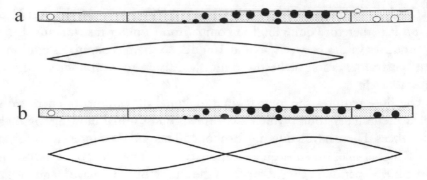

Fig. 7.8 Sketches of transverse flute with tone and register holes: **a** open (○) and closed (●) holes determining the length of a standing wave. **b** All holes closed except one register hole (in red). Corresponding standing waves are drawn below the flute bodies

Opening the tone holes from the far end makes the pressure node move closer up the bore—like making the pipe shorter. On the Böhm flute, each opened tone hole raises the pitch by a half tone. It is like a short circuit to the outer air, as if the flute were sawn off near the location of the tone hole.

Some special techniques mostly for professional flutists are cross-fingering and lipping. The former is for producing certain chromatic tones by closing one or more holes below an open one. The latter may lower the pitch by changing the chin position or rolling the embouchure hole towards them or away, thus changing the jet geometry.

7.2.4 Flutes from Various Areas

Other types of flutes have been invented, developed and played all over the globe (see Chap. 1). Below a few special flutes are briefly introduced.

- **Ney**
  The ney is an end-blown flute that figures prominently in Middle Eastern music. It has been played continuously for 4,500–5,000 years, making it one of the oldest musical instruments still in use.
- **Xun**
  The xun is a globular vessel flute from China made of baked clay or bone, similar to the ocarina (see Fig. 6.6d). It is one of the oldest Chinese instruments and has been in use for approximately seven thousand years.
- **Dizi or Xindi**
  The dizi is a Chinese transverse flute widely played in many genres of folk music, Chinese opera, as well as the modern Chinese orchestra. Most dizi are made of bamboo, some from jade.

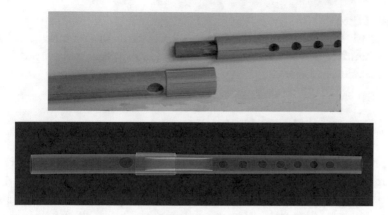

Fig. 7.9 Top: replica of a Nohkan, before joining two parts, bottom: X-ray photograph of Nohkan, reprint permission from Tokyo National Research Institute for Cultural Properties [9]

- **Nohkan**

  The Nohkan is a high pitched Japanese bamboo transverse flute. It is commonly used in traditional Noh theatre.[1] A very thin piece of bamboo is inserted between the mouth opening and the first finger hole, as shown in Fig. 7.9, top. An X-ray photograph reveals that the inner diameter of this part is narrower than that of the other part (Fig. 7.9, bottom). This narrow part is called "throat". High pitch tones are created via a Venturi effect, which is the reduction in fluid pressure that results when a fluid flows through a constricted section (or choke) of a pipe [8]. Tuning does not really work: The interval between normal tones and overblown tones is smaller than an octave. Extreme high pitch tones are believed to call dead persons or gods.

- **Bansuri**

  This bamboo flute is an important instrument in Indian classical music, and developed independently of the Western flute. Every bansuri has a specific key and tonal center, depending on the length, inner diameter, and the relative size and placement of the finger holes.

- **Gasba**

  The gasba or tamja is a musical style based on a wind instrument of the same name (gasba means "reed" in the Berber language), which is widespread in Tunisia and Morocco. This instrument produces a hoarse sound and forms the basis of pastoral music.

---

[1] Noh is a major form of classical Japanese dance-drama that has been performed since the 14th century.

- **Irish Flute**
  The Irish flute is a conical-bore, wooden flute of the type favored by classical flautists of the early 19th century. This flute is still played in every countryside in Ireland.

7.3   Double-Reed Instruments

After presenting the recorder and the transverse flute, we proceed to the reed instruments, sound generation of which is based on the vibration of reeds in an air flow. Reed instruments are the most often played woodwinds in modern European music. In orchestral music there are the oboe, the English horn, the clarinet, the bassoon and others (see in Fig. 7.10).

Reeds are tall grasses that grow in large groups in swamps, shallow water or on wet and soft ground. They have strong, hollow stems that are common materials for making mats or baskets. As small pieces of cane they are often inserted into the tubes of a woodwind instrument, which sets in vibration the air column inside the tube and makes a sound. As mentioned in Sect. 7.1, there are single-reed and double-reed instruments.

We start with double-reed instruments. The oboe, the English horn and the bassoon belong to these instruments. A double reed with a mouthpiece for an oboe is illustrated in Fig. 7.12.

When blowing an oboe or a bassoon, air is directed into a narrow slit between two tongues of reed, whereby this slit opens and closes alternately by vibrations of the reeds. This resembles the work of a pressure controlled valve transforming

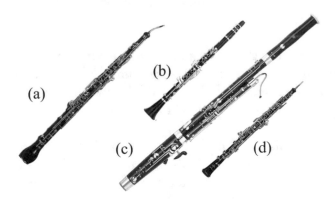

Fig. 7.10 Reed instruments: **a** English horn (CC BY-SA 3.0 by Hustvedt), **b** clarinet (CC BY-SA 4.0 by Yamaha Corporation), **c** bassoon (CC BY-SA 4.0 by Yamaha Corporation) and **d** oboe (public domain by Gerald G)

the air stream from the lungs of the player into air pulses which supply the air column in the tube with vibrational energy (see also Sect. 7.5.1).

After all, the tone is mainly determined by length and form of the inner bore of the tube, which is conical or cylindrical.

If it would be only so, then we would hear only overtone series like for the flute. But our life is not so simple. The double-reed instruments create more complicated sound. In addition to the overtone series "formants" contribute to the timbre. But what is a formant? Isn't it something that has to do with the human voice? Yes, a formant is the broad spectral maximum caused by an acoustic resonance of the vocal tract (refer to Sect. 8.3.3), where a pressure controlled valve mechanism works. In this sense a double-reed mouthpiece may function similar to the vocal tract, resulting in broad spectra like formants in the human voice. The location of such peaks is independent of the pitch played, but can depend on blowing strength and amount. Contribution of the formant to the timbre depends on the played pitch and its amplitude (strength).

J. Meyer reports the formant location of double-reed instruments compared with the human voice, as shown in Fig. 7.11 [10].

7.3.1 The Oboe

The name "oboe" is derived from French "hautbois" (= high wood), its spelling derived from Italian oboè. Its predecessor was the shawm, a double-reed aerophone made in Europe from the 12th century until today (Fig. 7.14 b). Double-reed instruments similar to the shawm were long present in Southern and East

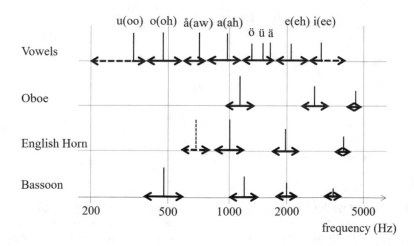

Fig. 7.11 Formant location of oboe, English horn and bassoon in comparison with the human voice (sung by professional singers). Data from Fig. 3.10 of [10]

Europe, for instance, the aulos of ancient Greece, the Etruscan subulo of Byzantine and the Roman Tibia. Furthermore, the Persian sorna and the Armenian duduk (Fig. 1.15). After having been removed from courtly music, the shawms are still played in folk music of the Mediterranean and in Brittany, as well as in oriental music.

7.3.1.1 Structure and Function

The oboe has been developed in the mid-17th century, usually made of special wood (grenadilla or African blackwood, also rosewood, palisander, ebony). It is 65 cm long, with a corpus consisting of three parts. Its tonal range is from B3 to A6.

As mentioned above, the double reed has a significant effect on the sound—its subtle manipulation has enormous influence on timbre and dynamics. Thus, most professional oboists scrape their own reeds to suit their individual needs. Before use it is put into water for a while.

Furthermore, playing the oboe involves complicated mechanics based on metal keys made of silver (or similar easily vibrating material), a conical bore and a flared bell as illustrated in Fig. 7.12. With its key mechanics the oboe belongs to the most complex wind instruments. The tone holes are sealed by beds of fish skin or cork.

Sound is produced by blowing into the reed with properly curved lips at a sufficient air pressure, causing it to vibrate with standing waves in the air column along the bore. Its characteristic tones are versatile, "bright" and clearly

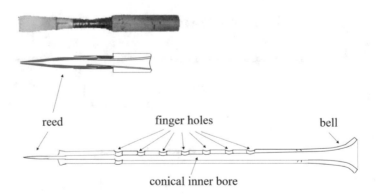

Fig. 7.12 Upper part: picture (CC BY-SA 3.0 by Aquazer) and graphic sketch of the mouthpiece: oboe reed with tube; the part between reed and tube is wrapped round with fish skin. Lower part: corpus with three sections, the double reed at the head, the inner bore with finger holes, and the bell at the foot

audible. Due to its expressive voice (pronounced higher formants) and comparatively penetrating sound the oboe is easy to hear over other instruments in large ensembles and is, therefore, widely recognized as the instrument that tunes the orchestra with its distinctive A4.

7.3.1.2 Sound Spectra and Dynamics

The oboe has a totally different tonal character than the flute. The role of the formants in the oboe sound is huge. When the 233 Hz (the fundamental for B♭3) is played, this tone is not well heard but, instead, the tones 1100 Hz (the principal formant) are heard as the strongest. This principal formant leads to a basic tone color similar to the vowel "ah" (see Fig. 7.11). The tone color towards a bright "oh" which comes from a subformant 570 Hz can be heard, too. Further, there are two secondary maxima at 2700 Hz which create the sound between "eh" and "i(ee)" and 4500 Hz lending a pronounced brightness.

From B5 (987 Hz) on the fundamental dominates, and thus the vowel-like tone color gives way, resulting in a hard and less expressive timbre. Therefore, in some compositions, an exchange between the part of the flute and the oboe may take place.

The "attack" of the oboe is very precise and clear, because initial transients are extraordinarily short without noisy contributions. Therefore, oboes are especially suited for a precise staccato with pearlescent clarity. The initial transient is shorter than 40 ms for the lowest note, and is lowered to less than 20 ms with increasing pitch. However, when it is necessary (for example, playing "cantabile"), the initial transient can be stretched to 100 ms. Such a soft attack would reach similar values as the violin for a sharp one.

The decay time of the oboe is also short: in the order of 0.1 ms [10].

7.3.2 The English Horn (Cor Anglais)

It is the alto instrument of the oboe family with a length of about 1.5 times that of an oboe (see Fig. 7.10). As a transposing instrument[2] it is pitched in F, a perfect fifth lower than the oboe, and its tonal range is from B3 to E6 (G6). On its top there is an S-bowed element connecting the reed with the instrument's corpus. The cor ends with a pear-shaped bell (German: "Liebesfuß"), which as a sound absorber creates a characteristic, warm, elegiac and less penetrating timbre.

---

[2]A transposing instrument is a musical instrument whose music is recorded in staff notation at a pitch different from the pitch that actually sounds (concert pitch).

Structure and etymology are quite interesting: the instrument originated in Silesia about 1720, when a bulb bell was fitted to an oboe da caccia-type body. They say that this resembled the horns played by angels in religious paintings of the Middle Ages—so the name "engellisches Horn = angelic horn" came about. On the other hand, the S or angle-shaped connecting tube between the reed and the corpus could have been a source to use the French term "anglé" for this particular feature. Thus, the instrument was called "cor anglé" with a language transition to "cor anglais = English horn" [11]. Who knows?

This instrument is famous for its musical intuition and wonderful romantic sound. Many orchestral masterworks listed below rely on its unique timbre.

G. Donizetti—Concertino for Cor Anglais and Small Orchestra
H. Berlioz—Symphonie Fantastique
A. Dvořák—Symphony No. 9 "From the New World" (2nd movement)
M. Mussorgsky—Il Vecchio Castello, from Pictures at an Exhibition (Orchestral version by M. Ravel)
H. Villa-Lobos—Choros No. 6
P. Hindemith—Sonata for Cor Anglais and Piano
B. Bartók—Concert for Orchestra

Other members of the oboe family (not often appearing in modern orchestras) are oboe d'amore (a third lower than the oboe), oboe da caccia (predecessor of the English horn), musette (a forth above the oboe) and heckelphone/ baritone oboe (one octave lower than the oboe).

7.3.3   The Bassoon

Among the family of woodwinds the deep-sounding bassoon plays the role of the grandfather, like in Prokofiev's symphonic fairy tale "Peter and the Wolf" (see Sect. 4.4). It is known for its distinctive tone color, wide range, variety of character and agility. The name comes from French bassoon. Its Italian name is fagotto (German: Fagott), probably meaning in old French a bundle of sticks.

This quite long "tubular" instrument (overall height 1.34 m, sounding length 2.54 m inclusive the S-bow) has a rather complex structure, as depicted in the following sketch (Fig. 7.13) [12]. The bell extends upward, the bass (or long) joint connects the bell and the boot, the wing joint extends from boot to bocal, the bocal (or crook) attaches the wing joint to a reed. This bocal is a curved tube, which connects the reed to the rest of the instrument. It is employed by inserting its cork end into the instrument.

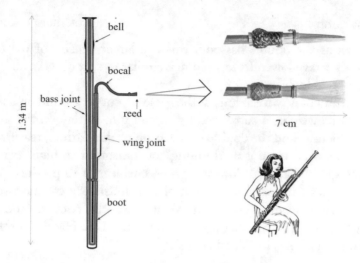

Fig. 7.13 Left: structure of a bassoon (CC BY-SA 3.0 by Gmaxwell), right: a reed (enlarged) from two different view points (GFDL 1.2 photo by Gregory Maxwell, illustration: public domain by Pearson Scott Foresman)

7.3.3.1 Sound Spectra and Dynamics

As seen from Fig. 7.11, the tone color is very similar to the formant for the vowel "o(oh)" which characterizes tones up to C4, whereas for the 2nd partial, the tone color resembles an "a(ah)". In still higher regions one perceives nasal components.

Although the bassoon is one of the deepest-sounding instruments, there is a high frequency contribution of the overtone series (12,000 Hz) which determines the tone, when it is played $f\!f$. At somewhat lower register the densely packed overtones assume a noisy character, which results in a certain tonal hardness. In contrast, at lower loudness levels, the principal formant is shifted towards a darker tone color. For pp the power spectrum 600 Hz drops drastically, so that the tone becomes rounder and more damped.

Despite of the low pitch, the attack of the bassoon is very precise, because the overtones in the middle and high frequency range have very short initial transients. And like for all wind instruments the decay processes are also short with a decay time around 0.1 s for the higher notes and up to 0.4 s for the lowest ones [10].

Contrabassoon

Also known as the double bassoon, this is a larger version of the bassoon, sounding an octave lower. Its technique is similar to that of its smaller cousin, with a few notable differences.

The instrument is twice as long as the bassoon, curves around on itself twice and, due to its weight, is supported by an end pin. It plays in the same bass register as the tuba and the double bass version of the clarinet, reaching down to $Bb0$. In its tone there is a "thinning" of the sound in high register, but a booming quality arises in lowest register, enabling it to produce powerful contrabass tones when desired. In symphonic music the contrabassoon is a featured instrument, most notably in Maurice Ravel's "Mother Goose Suite" and at the opening of the Piano Concert for the Left Hand. Gustav Holst composed several solos in "The Planets".

7.3.4   Other Double-Reed Instruments

There are many other double-reed instruments.

The **hne** (Fig. 7.14 a) is a conical shawm of double reed used in the music of Myanmar. It has a sextuple reed made from young leafs of a sugar palm tree, which is soaked for six months. Its body is made of wood with a conical bore and seven finger holes. It has a flaring metal bell and a loud tone used in an ensemble together with harps, xylophone, gongs and contrasting with the sound of drums.

Its provenance is the Persian area and, around the year 1500, it came through Arabia to East Asia including Malaysia and Thailand. Its tonal range is two octaves. Due to its thin or sharp penetrating sound, it is preferentially used in large orchestras. Similar instruments have been in use in countries like Northern India (tangmuri), Tibet (gyaling) or Nepal (mvali).

The **shawm** (Fig. 7.14 b) was made in Europe from the 12th century to the present days. It may have originated in the Eastern Mediterranean around the time of the crusades and then gained great popularity in the Renaissance period. Its double reed is made from the same "arundo donax" cane used for oboes and bassoons.

Furthermore, there are the **kèn bầu**—used in the former royal court music of Vietnam; the **nadaswaram** from South India used as a traditional classical instrument, played in pairs and accompanied by two drums called thavil; the **algaita** from West Africa, characterized by a large, trumpet-like bell; the **hichiriki** as a Japanese fue (flute) used as one of two main melodic instruments

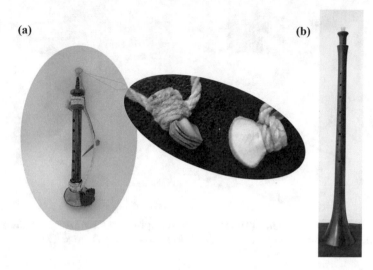

(a)   (b)

Fig. 7.14 Double-reed instruments. **a** Hne from Myanmar, ancient type of an oboe, right: enlarged pictures of the reed, consisting of 6 lamellae (photos by KT), **b** shawm (CC BY-SA 3.0 by Jan Klimeš)

in Japanese imperial court music; and the **cromorne**, French reed instrument of uncertain identity, used in the early Baroque period in French court music (name not to be confused with the similar-sounding name crumhorn).

7.4   Single-Reed Instruments

Single-reed instruments are also widely spread in many countries. Beyond the clarinet, there are ancient instruments of this kind already documented in Egypt, as well as in the Middle East, in Greece, and in the Roman Empire. In early times such instruments used idioglottal reeds, where the vibrating reed is a tongue cut and shaped on the tube of a cane.

Later, single-reed instruments started using heteroglottal reeds, where a reed is cut, separated from the tube of cane, and attached to a mouthpiece of some sort. An idioglottal instrument poses almost no dynamic range. Small changes of blowing pressure result in alteration of tonal frequency, large ones cause overblowing. This is different for the heteroglottal case, where broader reed blades and tube diameters enhance the dynamic range.

The modern single-reed instruments like the clarinet are characterized by just one vibrating reed attached to a solid beaked-shaped piece of wood or hardened rubber. Upon a stream of air blown into the mouthpiece, the reed fixed on

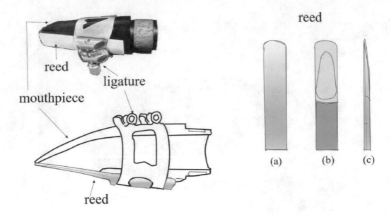

Fig. 7.15 Left: structure of mouthpiece of clarinet (photo: CC BY-SA 3.0 by James Eaton-Lee Njan), right: **a** reed: outside, **b** inside, **c** side view; size: 8.1 cm x 1.7 cm

it starts to vibrate. An example of the single reed attached to a mouthpiece is shown in Fig. 7.15.

7.4.1 The Clarinet

The name "clarinet" (small clarion) stems from the fact that in its high registers it sounds like a high clarin-trumpet (a Baroque trumpet for higher registers), taking in the 18th century a part of its role. The modern clarinet is an important member of symphony orchestras.

A significant difference between a clarinet and an oboe or a bassoon is the relationship of the tone to the length of the tube. As seen in Fig. 7.10b and d, the body length of the clarinet tube and that of the oboe tube are almost identical. Both an oboe and a B♭ clarinet have 70 cm of body length. The tube lengths of these two instruments also do not differ so much. From this fact we expect that the lowest pitch (the fundamental) of both instruments should be almost the same. However, the lowest tone of the oboe is B3, while that of the B♭ clarinet is D3. The clarinet creates much lower tones. Why?

In Sect. 6.2.5, we discussed standing waves in an open tube (both ends are open) and a closed tube (one end is open, and the other end is closed). The wavelength of the standing wave in a closed tube is twice longer than that in an open tube. As a consequence, it sounds one octave lower than a tube with two open ends. Therefore, we could suppose that a clarinet behaves like a open system at its lower end, but the mouthpiece end is considered close.

In fact, the clarinet is open at the far end or bell. But it is (almost) closed at the other end. For a sound wave the tiny aperture between reed and mouthpiece—

a much smaller cross section than the bore of the instrument—is enough to cause a reflection almost like that from a completely closed end.

Consequently, the total pressure at the open end of the pipe must be approximately atmospheric pressure. The mouthpiece end, on the other hand, can have a maximum variation in pressure. Now, the distance between a zero and a maximum of a sine wave is one quarter of a wavelength. So, the longest standing wave that can satisfy these conditions is one that has a wavelength four times the length of the instrument.

As illustrated in Fig. 7.16, for the basic tone, only a standing wave with a quarter wavelength is contained in the tube. When initiating the first higher harmonics by overblowing, the clarinet produces the twelfth (duodecimo, i.e. from 1/4 to 3/4 of the wavelength), whereas a flute would radiate the octave. The next step of overblowing comprises 2 octaves and a third (i.e. 5/4 of the wavelength). In effect, the clarinet possesses thus a large tonal range up to 4 octaves.

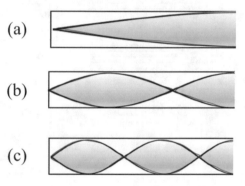

Fig. 7.16 Standing waves for fundamental (**a**) and the two first harmonics (**b**), (**c**) in a tube closed on one side

The clarinet presents a typical example for sounds, for which the odd partials outweigh the even ones. But they do not preserve these characteristics over the entire pitch range. They change their spectral composition in the region of the higher register in favor of the even tone contributions. In the low octave, up to the 15th partial, the second and fourth partials are particularly weak. Further up in frequency the odd and even contributions are equally developed. Above G♯5 the fundamental dominates in a strong way. This is associated with a steadily decreasing overtone series giving a "whole round substance". A formant between 3 and 4 kHz (vowel color "i(ee)") adds to tone brilliance and clarity of sound [10].

Attacks with the clarinet can be very clear and precise. The lowest notes, even the fundamental, reach their final strength within a few vibrational periods. The initial transient is completed 15–20 ms after the attack. The decay time, even with low notes, is not longer than 0.2 s and drops to about 0.1 s in the upper register.

Notation and Sound

The Bb clarinet is a transposing instrument. When the note C occurs in a score that is especially written for a Bb clarinet (the upper note), the instrument will actually sound as Bb (the lower note), hence the name of the instrument. The very same fingering when played on an A clarinet will sound as A, and in a score for A clarinet, an A will be written as a C.

notation

sound

Bb clarinet   A clarinet

Clarinet Family

The clarinets form a big family. Members are, from the higher pitch range to the lower range, the Eb clarinet (sopranino), C clarinet (soprano), Bb clarinet (soprano), A clarinet (soprano), basset clarinet (mostly in A), basset horn (in F), alto clarinet (mostly in Eb), bass clarinet (in Bb or A), contra-alto clarinet (in Eb) and contrabass (double-bass) clarinet. As extended members there are the EEEb (octocontra-alto) and BBBb (octocontrabass) clarinets.

Among these ten members the Bb clarinet is most commonly used in orchestras. Its pitch range is D3—Bb6. This clarinet feels lonely, however, without a support of his elder sister, the A clarinet (pitch range: Db3 - A6). Therefore, most of the time the Bb clarinet player brings an A clarinet along and puts it on his side. The Bb clarinet has, due to its richer harmonic spectrum, a more glamorous and powerful charisma, whereas for the A-clarinet the darker and cantabile character is more pronounced. Passages in a flat key are easier performed by the Bb-instrument, whereas the A-instrument is better suited for sharp keys.

Fig. 7.17 **a** Bass clarinet in B♭ (length ≈ 130 cm) (CC BY-SA 4.0 by Yamaha Corporation), **a'** shows how big the bass clarinet is for a woman of the standard size, **b** tenor saxophone in B♭ (length ≈71 cm) (CC BY-SA 4.0 by Yamaha Corporation)

The bass clarinet (Fig. 7.17a) is also relatively often used. This instrument is usually pitched in B♭ and plays notes an octave below the soprano B♭ clarinet, i.e. with an extended range from D2 to B♭5. Playing this instrument requires support with an end-pin introduced to give the instrument greater stability.

7.4.2 The Saxophone

This instrument was invented by the Belgian instrument maker Adolphe Sax in the early 1840s (shown in Fig. 7.17b). It is a woodwind instrument usually made of brass and played with a single-reed mouthpiece. Sax constructed several series of instruments ranked by pitch, among which the series pitched in B♭ and E♭ soon became dominant and enjoys highest popularity until today. The pitch is controlled by keys with shallow cups in which are fastened leather pads that seal tone holes, thus determining the resonant length and thereby frequency of the air column in the bore. The player covers or uncovers the holes with help of levers. Other than with a clarinet, the saxophone overblows the octave like flute or oboe and not the twelfth, owing to the conical shape of the bore.

7.4.3 Other Single-Reed Instruments

The **chalumeau** (Fig. 7.18a), a historic instrument, is considered as a predecessor of the modern-day clarinet, frequently played in the late Baroque and early classical eras. First mentioned during the 1630 s, it may have been in use as early as in the twelfth century. Since in early constructions the instrument had a quite limited range of only 12 notes, multiple sizes were produced ranging from bass to soprano. The chalumeau has an intimate, cantabile-like quality— as opposed to the trumpet sound of the Baroque clarinet, and similar to the sound of speaking.

The **alboka** (Fig. 7.18b) (Pays Basque) has its name from Arabic al-boc, meaning horn. It is held with both hands during playing. The **mijwiz** (Fig. 7.18c) (from Egypt, Middle East and Albania) means in Arabic "dual", often used in belly dance. It is operated with circular breathing, not unlike for a bagpipe, but with the air reservoir in the mouth.

There are the **mock trumpet**, popular during the second half of the seventeenth century, especially in England; the **birbyné** from Lithuania, with its body made of maple or pea; the **sneng** from Cambodia, made from the horn of an ox or a water buffalo; the **launedda** as a set of three tubes from Sardinia— these have different length; the longest one is a "bordun" tube with one hole for a deep tone for accompaniment; the **pibgorn** from Wales—a horn pipe with a reed covered by a wind capsule; the **pku** from Armenia—with an open cone made of bull horn; the **zhaleika** from Russia—a sheperd's instrument, known as "folk clarinet"; and the **sipsi** from Turkey—made of bone, wood, or reed and used in instrumental folk music.

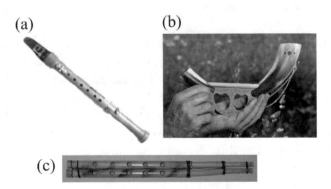

(a)   (b)

(c)

Fig. 7.18 Single reed instruments. **a** Chalumeau (CC BY-SA 4.0 by Maui2903), **b** Alboka of Pays Basque (CC BY 3.0 es by Mikel Arrazola, **c** Mijwiz of Egypt, Middle East, Albania (CC BY-SA 3.0 by Sönke Kraft aka Arnulf zu Linden)

7.5 Brass Instruments

Similar to the flutes brass instruments have old traditions. Since ancient time it is known that a tube of conical form functions as an amplifier (like megaphones). In the old Egypt armies one used kind of natural conical instruments. Later some of them became horns, and others trumpets [14, 15].

As described in Chap. 1, most of the instruments consist of a sound generator and resonator (Fig. 1.4). As generators brass instruments are somewhat special, because vibration is created by lips. It means that a human organ is an intimate part of the instrument. Air is blown constantly through slightly stressed lips: When the air flow reaches a certain velocity, the lip tissue starts vibrating due to the muscle resistance. This mechanism acts in a similar way on the vocal cord in our speech organs (see Sect. 8.3.2) and is called a pressure controlled valve (in German "Polsterpfeife").

Note that the classification of brass instruments is independent of the materials. If we try to define the brass instrument exactly, we can say "the lip-vibration instrument". For example, an alpine horn is made of wood, but it is categorized as a brass instrument, because sound is created by lip vibration. On the other hand, a saxophone is made of metal, but it is a woodwind instrument, because sound is created by reed vibration.

7.5.1 Physics of Sound Creation

As said above, a brass instrument functions as a pressure controlled valve. The closed lips, which are laterally strained, are opened by the air pressure, so that the pressure difference between the inside and outside of the lips is suddenly balanced. At the same time the area close to the pathway of the air develops low pressure and, therefore, the lips are closed. Repeating these processes causes a vibration, as illustrated in Fig. 7.19.

Such vibrations are similar to what happens in a linear mass-spring oscillator, because lips are springy, and they have mass. However, in the case of the lips, the oscillation occurs in two dimension.

When an instrument is present at the outside of the lips (usually there is a mouthpiece of a cup or funnel formed at this place), this leads to a standing wave in the bore of the instrument. Figure 7.20 shows standing waves of the first, second and third natural tones (refer to Sect. 4.3.2.4 in Chap. 4) in a conical tube, one side of which is open. Due to the conical form distances to the next maximum are not equal and, therefore, these distances have no direct relationship to the emitted tones.

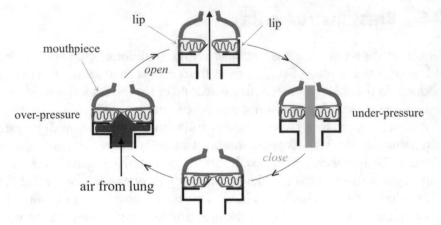

Fig. 7.19 Principle of a pressure controlled valve. Area marked in red is over-pressured, and in blue under-pressured

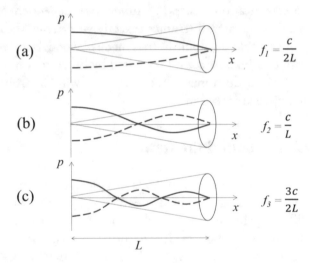

Fig. 7.20 First three normal modes of vibration for a conical tube. Red curves are standing pressure waves

In order to obtain the best possible sound (as many harmonics as possible), the lip frequency should be congruently identical to the natural tones. If the initial frequency of the lips is higher or lower than the natural tone, the standing wave becomes shorter or longer, resulting in a higher or lower "wrong" tone.

If the tube is approximated to be conical, its narrow end is closed with the players mouth (although some parts of the tube are cylindric). Therefore, the lowest tone has a wavelength of about 2 times longer than the length of the instrument (Fig. 7.20a). The next higher eigenfrequency (the second natural tone) has a wavelength equal to the length of the instrument, which is one octave higher than the first natural tone (Fig. 7.20b). Here the yielded eigenfrequencies form a well known natural harmonic series. Adjusting the frequency by the acceleration of velocity of the air flow between lips can cause an overblow, which results in the integral multiplication of the fundamental frequency.

The highest playable tone depends on the mouthpiece. A small and narrow cup can create higher tones than a large and wide cup. However, these tones are especially sharp.

The form of the bell is also important. Usually, the bell is not a simple extension of the conical tube, but has the form of an exponential funnel. Such a funnel transforms the wave impedance from the very large value of the tube to the small inferior value of the air, resulting in an effective radiation of the sound energy into the surroundings. Only the remaining energy is reflected back to the tube for forming the standing wave. The funnel affects not only the sound color, but also the interval distance of the natural tones, because the gradual change of the diameter arises in the change of the acoustic length of the tube. The reflection point of the standing wave outside of the funnel can be shifted further. However, this phenomenon is not yet well studied, since other factors (for example, the form of the inside of the player's mouth, position of the tongue) are involved.

7.5.2 The Horn

7.5.2.1 Natural Horn

A natural horn is a common definition of brass instruments having a shape like a French horn (Fig. 7.21b, c). It is blown according to the principle of the pressure controlled valve, but possesses neither finger holes nor valves, nor a telescope mechanism like a trombone. This means that a natural horn cannot create any tones except natural tones, which are determined by the total length of the blown tubes. An alpine horn, Baroque horn, hunting horn, lur, parforce horn, post horn (Fig. 7.21b), Russian horn, fire horn, signal horn (see Fig. 7.21a) belong to the natural horns.

Since the 18th century natural horns have been used in orchestras. At the beginning they could create only 12 natural tones by using lips. Later, players put a hand into the bell opening, so that more tones could be created. Their timbre is quite different from the modern horns with valves. Various pitches can be played due to an extension of the tube. See below one example of using a natural horn in an orchestra piece by Carl Maria von Weber:

Cadenza ending of Concertino for (natural) horn and orchestra by Carl Maria von Weber

7.5.2.2 Modern Horn with Valves

Since in 1818 rotary valves have been applied to horns (see Fig. 7.21c) by Heinrich Stölzel and Friedrich Blühmel, various pitches can be played without changing the length of the tube or putting a hand into the bell. At the beginning only two valves were built in. Later a third or even fourth valve were added in order to have more flexibility. In France piston valves were developed.

However, orchestra people strongly rejected to use such modern horns for works of Carl Maria von Weber. One of the first compositions which was

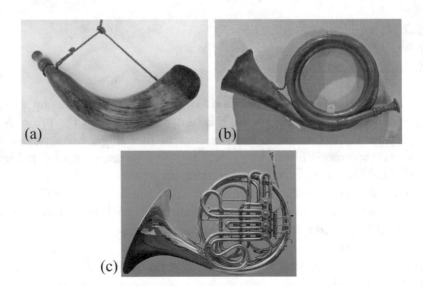

Fig. 7.21 a Signal horn made from a cow horn with wooden reed, in the Museum for Hamburger History (CC BY-SA 3.0 Photo by Bullenwächter), **b** post horn (19th century); exhibition in the Spandau citadel, Spandau, Germany (CC BY-SA 3.0 by JoJan), **c** a modern (double) horn with rotary valves (CC BY 2.5 by BenP)

played with a horn with valves is the Nocturne of "Ein Sommernachtstraum" (A Midsummer Night's Dream) by Felix Mendelssohn Bartholdy.

Initially valve horns were tuned in F key, which, although easier to play accordingly, has a less desirable sound in the mid and low register. Soon after, for better and higher tones, a shorter horn in B key was constructed. An F horn has F2 (87.3 Hz) as the lowest natural tone. The theoretical length of the tube can be calculated for $T = 20\,°C$ as

$$L = c/F = 343 \text{ (m/s)} / 87.3 \text{ (s)} = 3.93 \text{ m}$$

with 343 m/s sound velocity. A B horn has Bb2 (116.6 Hz) as the lowest natural tone and the length of the tube is 2.94 m for $T = 20\,°C$. However, the length of a real horn is 3.6 m for the F horn and 2.7 m for the B horn. The discrepancies may be due to the form of the tube: it is not cylindrical but conical.

Similar to the clarinet the horn is a transposing instrument. Its sounds are deeper than the written notes in all tunings. The written notes of the modern horn (the F horn) are a fifth higher than the corresponding sounds.

7.5.2.3 Double Horn

In 1897 Eduard Kruspe und Bartholomäus Geisig constructed a double horn. The length of the F horn tube can be extended by about 100 cm (integrated inside the main tube) with help of a main switching valve, so that the F and B tunings can be realized in one instrument. A triple horn F/B/high F was also constructed.

Coffee Break

Horn players put their right hand into the bell, when they play. Why? There are several reasons for this. At first, a rather heavy instrument should be supported also with the right hand: it is more practical to put the hand into a wide opening than to hold a thick tube from outside. For the second reason they can make fine tuning by changing the position of the hand inside. It seems that horn players cannot forget their beautiful time with natural horns. Third, they can create very sharp tones when the hand is put very deeply. Finally, the hidden hand can be a secret tool for communication: "When should I start again?" "Three bars later..."

7.5.3 Trumpet, Trombone, and Tuba

Beyond horns, trumpets, trombones and tubas are also often used brass instruments in orchestras. Figure 7.22 shows these three instruments in their relative sizes (the length of the tuba is 874 mm). Similar to the horn, sound is created by the vibration of lips due to the principle of a pressure controlled valve.

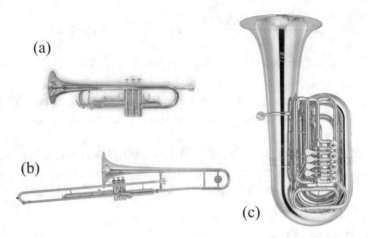

Fig. 7.22 a Trumpet (CC BY-SA 3.0 by PJ), **b** trombone (CC BY-SA 4.0 by Yamaha Corporation), **c** tuba (CC BY-SA 4.0 by Yamaha Corporation)

Their bores are approximately conical, the narrow end of which is closed with the players mouth. Therefore, comparable to the case of the horn, the lowest tone has a wavelength of about 2 times longer than the length of the instrument, and the second natural tone has a wavelength equal to this length, which is one octave higher than the first natural tone (see Fig. 7.20).

Trumpet

The trumpet (Fig. 7.22a) is the highest-pitched musical instrument in the brass family. And it is played not only in classic but also in jazz ensembles. The sound is colorful and brilliant.

The natural trumpet has neither valves nor pistons, and therefore, it can create only natural harmonics. This did not matter for military march. But it was not good enough for playing music. Nevertheless, in the Baroque period trumpets of better quality (for, example, smooth surface of the tube inside) and high techniques of the players (sophisticated use of lips, tongue, teeth) enabled to perform Bach's and Händel's works.

Only after valves and pistons were invented by Stölzel and Blühmel in the 19th century for horns, valves were applied to trumpets, becoming acceptable as musical instruments in orchestras. Now the Bb and C trumpets are most commonly used. Pitches can be controlled with either rotary or piston valves.

The mouthpiece plays an important role for timbre and quality of sound. It affects also the playability for players. A circular rim provides a comfortable environment for the vibration of lips. As shown in Fig. 7.23, the cup behind the rim channels the air into a much smaller opening (the backbore) that tapers out slightly to match the diameter of the pipe of the trumpet. When the cup is wide and deep, the sound is dark and soft. On the other hand, a shallow cup creates sharp and bright tones. Usually players have various mouthpieces, which they use depending on the requirements of musical pieces and performances.

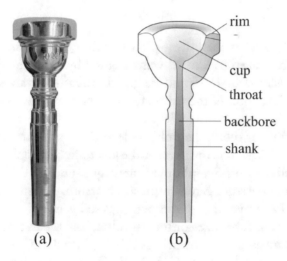

rim
cup
throat
backbore
shank

(a)   (b)

Fig. 7.23 **a** Mouthpiece of a trumpet (CC BY 3.0 by Eusebius (Guillaume Piolle)), **b** its section and name of parts (CC BY-SA 2.5 by WarX (Artur Jan Fijalkowski))

Trombone

The root of a trombone is the same as that of the trumpet/horn. However, different from other brass instruments with valves, trombones have a telescoping slide system, which varies the length of the instrument to change the pitch as shown in Fig. 7.22b. (Note that modern trombones often possess valves for lower pitches.) When the length of the air column is extended by sliding, the pitch is lowered. Because of the continuous change of the length, trombones can change pitches continuously. Therefore, it is possible for trombones to play microtones (see Sect. 3.2.6) like 1/4 or even 1/8 intervals. Vibrato with microtone intervals gives us a nice feeling of rolling.

Since the 15th century trombones were played widely in courts, churches and military groups. From the 18th century on, however, they were used mainly in a religious context, probably because of their glorious harmony. During the late Baroque period, Johann Sebastian Bach used trombones for his cantatas (for example, BWV 2, BWV 21 and BWV 38) and Georg Friedrich Händel used trombones in the Death March from Saul, Samson, and Israel in Egypt in an oratorio style.

Ludwig van Beethoven is one of the first composers who introduced trombones to the symphony orchestras. He used trombones in his Symphony No. 5 in C minor ("Fate Symphony"), Symphony No. 6 in F major ("Pastoral") and Symphony No. 9 ("Choral"). Since then trombones are important members in most of the symphonic works.

Tuba

The tuba (Fig. 7.22c) is the lowest-pitched musical instrument in the brass family. The tuba (called bass tuba) was invented by Wilhelm Wieprecht und Carl Wilhelm Moritz in 1835 and patented. Tubas had valves from the beginning. (The invention of the valves for aerophones were made in the 18th century.)

Before the time of tubas, the ophicleide and the serpent (see figure in the graybox in Sect. 6.1.4) took the part for the low tones. The ophicleide was used by Hector Berlioz in his Symphonie fantastique, and by Felix Mendelssohn Bartholdy in the overture Ein Sommernachtsraum. When Richard Wagner composed the Faust overture, the serpent was in his mind.

There are various tubas like contrabass tubas, bass tubas, tenor tubas,... The lowest pitched tubas are the contrabass tubas, pitched in C or B♭, the fundamental pitch of which is 32 Hz (C1), 29 Hz (B♭0), respectively.

Many composers wrote famous pieces using a tuba or tubas.

Examples are:

- Richard Wagner: Die Meistersinger von Nürnberg, Lohengrin
- Richard Strauss: Also sprach Zarathustra, Eine Alpensinfonie, Till Eulenspiegel, Ein Heldenleben
- Hector Berlioz: Symphonie fantastique
- Modest Mussorgsky (orchestra version by Maurice Ravel): Pictures at an Exhibition—Bydlo
- Sergei Prokofiev: Fifth Symphony
- Igor Stravinsky: The Rite of Spring, Petroushka

- George Gershwin: An American in Paris
- Gustav Holst: The Planets.

7.6 The Organ

The organ is the "Queen of instruments", says Wolfgang Amadeus Mozart. This composer, who created universal and globally appreciated music for virtually any instrument available at his time, did not leave behind many original works for organ, just several demonstrations of his virtuous playing on the organ and quite short musical ideas that are collected today in a series of pieces to be performed on the piano or on the organ (arranged church sonatas).

Nevertheless, we have to acknowledge the great history of this musical instrument, which is the biggest and most complex invented by mankind. The organ is a relatively old musical instrument dating from the time around 250 BC, when the water organ was invented. It was played in the Ancient Greek and Roman world. In Medieval times it spread from the Byzantine Empire to Western Europe, where it assumed a prominent place in the liturgy of the Catholic church. As a consequence, most organs in Europe, the Americas, and Australasia can be found now in Christian churches.

The number of pipes are a measure of the organ's size. To be remembered: a simple predecessor of these pipes is the pan flute (compare Fig. 6.6b), consisting of perhaps 20 tubular pipes, played by blowing across their upper opening. Nowadays, the largest organ world-wide is the one in the opera house of Sidney in Australia having 10,500 pipes and 5 manuals. Previously, in 1429, one had already constructed an organ with 2,500 pipes in Amiens, France.

During the centuries, organs of different style and size have been installed in numerous churches. A rather small chamber organ is shown in Fig. 7.24. Georg Philipp Telemann, Johann Sebastian Bach composed many pieces for organs. Along with the development of pianos the importance of the organ in composition decreased. Nevertheless, some impressive oeuvres for organ were composed by Felix Mendelssohn Bartholdy and César Franck.

We will focus here on the heart of this instrument which consists of a collection of organ pipes radiating the multifaceted sound which we adore when listening to this instrument, and we start with a very small musical aerophone, the whistle.

Fig. 7.24 Chamber organ by Pascoal Caetano Oldovini (1762), Meadows Museum, Dallas, Texas (Public domain, photo by Andreas Praefcke)

7.6.1   The Whistle

The whistle (or pipe) has been a sound producer since very ancient times. Made of bone or wood it has been used for thousands of years. It served as a signaling instrument for many purposes, mouth-operated or powered by pressured air or steam, for instance in hunting or transmitting important news or to keep the stroke of galley slaves. Or to signal orders to archers, aboard naval sailing vessels to issue commands. It can be taken as a basic element for the huge organs that function on the same principles as those operating for wind driven pipes like the recorder (see Sect. 7.2.2). We have learnt how the head of the recorder functions, which is nothing else than a whistle with a windway through which air is directed against the bladed edge of a labium. The air is broken there, starts vibrating and thus generates sound waves. A small attached resonance room— a short tube or some other oblong structure—supports a pure, or nearly pure, tone.

An especially intense sound is produced by the pea-whistle, first done so by J. Hudson in 1883 (see Fig. 7.25): if a pea is put into a closed whistle, sound may travel for a distance over a mile. A regularly occurring change of the cross-section of the resonance room due to the enclosed pea results in a corresponding, small change of pitch. Our ear perceives that as two simultaneous tones which interfere and are dissonant. This causes discomfort and enhances the signaling effect. Close to the ear the trill may reach the pain level. The

Fig. 7.25 A plastic pea whistle with trill effect (photo by KT)

"trill" effect is used in many situations: trains, ball games, police, military, ship wreckage, and more.

The aerophones "whistle" and "pipe" are both called "Pfeife" in German, indicating that the distinction between them is not always consistently made.

7.6.2 The Organ Pipe

The musical sound of an organ is produced by organ pipes. These are present in large number, each of them specialized for a particular pitch, timbre and dynamics. One distinguishes between flue (or labial) pipes with no moving parts, based solely on the vibration of air, in the same manner as a recorder, and reed pipes, where sound is produced by a beating reed (a curved piece of brass), using the same principle as for a clarinet, for instance. There are other less frequently used types like the diaphone pipe.

We focus here on the most common case: the labial pipes. They are generally made out of either hardwood or metal. Quite rarely glass, porcelain, or plastic may be found. The metal is usually lead alloyed with tin along with trace amounts of antimony and copper, which influences the characteristics of the resulting pipe. Shapes are cylindrical, conical or rectangular (Fig. 7.26).

The labial pipe functions similar to a flute: air is flown through an opening in the lower foot part; an air stream flows like a thin lamella through a narrow slit—the windway—from the lower to the upper labium. Thereby most of the pipe's cross-section is blocked by the languid. The air sheet above the sharp edge of the upper labium is periodically diverted, which gives rise to a sound wave in the upper part of the tube.

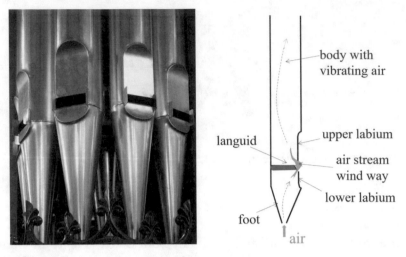

Fig. 7.26 Left: organ pipes with bronze patina on the mouth (public domain by Mego-denas, source Recadrage Photoshop), right: structural details

7.6.3 Wind System

The many pipes of an organ need "living air" to breathe, so the organ has a lung via pumping pressured air. For stable tones to develop a constant wind pressure must be generated [16]. In earlier times this was realized with help of calcants (from Latin calcare = to step), who were employed for service at bellows or other devices to furnish strong blasts of air thereby using hands and feet. For large organs up to 12 calcants had to work. Special techniques were involved to ensure constancy of the wind pressure. On the other hand, additional arrangements were introduced to create a tremolo sound (by the Tremulants). In fact, a whole technology was concerned with constructing efficient ways for collecting pressured wind that could be transferred to the reservoirs of an organ. The pressure values would have an influence on the tonal differentiation of some parts of the organ, There are quite interesting historical examples of a wind work realization, for instance the use of box bellows: Here a wooden box is loaded on its upper side with heavy stones and positioned into another fitting box.

At the age of 12 years SCM got a job to do together with his brother, Rainer (having similar weight): go to the village church, enter the church tower and assist their mother who on Sunday mornings played the organ on the gallery. Not in terms of music but as calcants to keep the air blowing through the organ pipes. What a burden! But they learned to enjoy their job, lifting heavy, stone-loaded plates up and down within the restricted space of the small church tower. Rainer's task was to step into a foothold on the upper stage and raise one stone-loaded plate up, while he himself was going down. So after having reached the bottom and leaving the foothold the plate would slowly move down to produce air blown into the registers of the organ. In between Rainer should climb up the ladder to the upper stage again. In parallel Stefan would do the same thing, but shifted in time by half a cycle. Finally there would be a (more or less) continuous flow of air to keep the organ alive. The quite intricate task was for the two to synchronize their action, which was difficult but also rewarding in case of success (see Fig. 7.27).

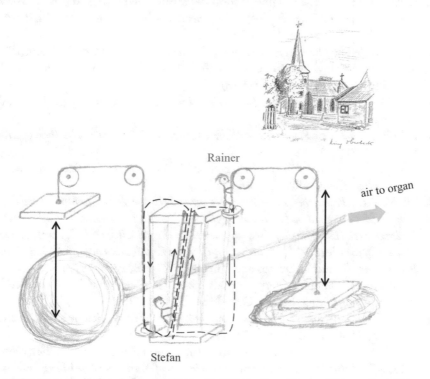

Rainer

air to organ

Stefan

Fig. 7.27 In a small church (top, right) built 1875 in the village of Westerlinde (close to Wolfenbüttel north of the Harz mountains), air for the pipe organ was created by two children. Drawing of the church by H. Oberbeck in 1952

SCM left the church tower together with his brother after the Sunday service, tired but with a good feeling.

Nowadays the mechanical constructions for wind blow are replaced by electrically driven devices: radial ventilators have taken over, for instance. The problem of undesired disturbances by motor noise and transmission through the air can be solved by placing the ventilator in a separated room.

Acknowledgments Authors thank Dr. Harald Kleine of University of New South Wales, Canberra for carrying out Mach-Zehnder experiments together with KT for visualizing air movement during playing a recorder. Bettina Geiger of the Dortmund Philharmonic Orchestra is acknowledged for offering us an opportunity of take pictures of her flute and piccolo.

References

1. M. Praetorius, *Syntagma Musicum II* (Elias Holwein, Wolfenbüttel, 1619)
2. H. Kleine, High-speed imaging of shock waves and their flow fields, in *The Micro-World Observed by Ultra High-Speed Cameras*, ed. by K. Tsuji (Springer International, Cham, 2018), pp. 127–155
3. J.W.S. Rayleigh, *The Theory of Sound*, vol. 1 (Macmillan and Co., London, 1877)
4. H. Levine, J. Schwinger, On the radiation of sound from an unflanged circular pipe. Phys. Rev. **73**, 383–406 (1948)
5. R. Le Roy, *Die Flöte - Geschichte, Spieltechnik, Lehrweise*, ed. by C. Dorgeuille (Bärenreiter, Kassel, 2002)
6. T. Böhm, *Über den Flötenbau und die neuesten Verbesserungen desselben* (Schott, Mainz, 1847)
7. J. Meyer, *Acoustics and the Performance of Music* (Springer Science and Business Media, Heidelberg, 2009), pp. 64–69
8. J.B. Venturi, *Recherches Experimentales sur le Principe de la Communication Laterale du Mouvement dans les Fluides appliqué a l'Explication de Differens Phenomènes Hydrauliques. (Experimental Investigations into the Principle of the Lateral Communication of the Movement in Fluids Applied to the Explanation of Different Hydraulic Phenomena)* (Houel et Ducros and Théophile Barrois, Paris, 1797)
9. I. Takakuwa, Manufacturing methods for nohkan and ryuteki as clarified by radiography. Res. Rep. Intang. Cult. Herit. **3**, 1–20 (2009)
10. J. Meyer, *Acoustics and the Performance of Music* (Springer Science and Business Media, Heidelberg, 2009), pp. 70–86
11. M. Finkelman, Oboe: III. Larger and Smaller European Oboes, 4. Tenor Oboes, (iv) English Horn, in *The New Grove Dictionary of Music and Musicians*, 2nd edn., ed. by S. Sadie and J. Tyrrell (Macmillan Publishers, London, 2001)
12. W. Jansen, *The Bassoon: Its History, Construction, Makers, Players, and Music* (Uitgeverij F. Knuf, Flanders, 1978)
13. F.G. Rendall, *The Clarinet - Some Notes Upon Its History and Construction* (Ernest Benn Limited, London, 1957)
14. A. Baines, *Brass Instruments: Their History and Development* (Dover Publications, Mineola, 2012)
15. P. Bate, *The Trumpet and The Trombone. An Outline of Their History, Development and Construction,* (Ernest Benn, London, 1966)
16. S.A. Elder, The mechanism of sound production in organ pipes. J. Acous. Soc. Am. **54**, 1554–1564 (1973)

8

Selected Instruments 2—Chordophones, Membranophones, Idiophones and Human Voice

8.1 String Instruments

Among the large number of musical instruments the group of string instruments plays one of the most important roles.

8.1.1 The Monochord

Let us start with a simple set-up, which is the monochord, an ancient device where a string is fixed at both ends and stretched over a sound box. One or more movable bridges are then manipulated to demonstrate mathematical relationships among the frequencies produced (Fig. 8.1).

The monochord is mentioned already in Sumerian writings and was supposedly "reinvented" by Pythagoras and other Greek music theoreticians whose studies on harmonic intervals have been described in Chap. 4. There are other early musical constructions involving one or many strings, like the harps in Egypt or the phorminx, a predecessor of the lyre and kithara in ancient Greece, already mentioned by Homer: these early instruments are shown in Figs. 1.14 and 4.1.

8.1.2 The Vibrating String

The string is a cylindrically formed element, very thin as compared to its length and soft for bending. One can fix its ends and exert tension (resilience). Then the string may start to vibrate: its radial displacement is mainly characterized

© Springer Nature Switzerland AG 2021
K. Tsuji et al., *Physics and Music*,
https://doi.org/10.1007/978-3-030-68676-5_8

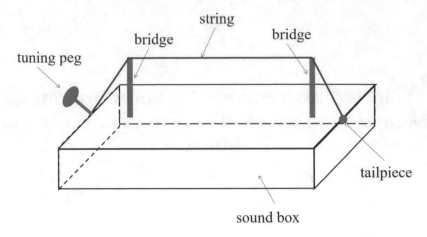

Fig. 8.1 Monochord

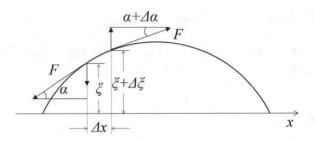

Fig. 8.2 Vibration of a string

by the effect against the axially effective force of the tension together with the elasticity and the mass of the string leading the system back to the initial state. Sound is produced by coupling to the surrounding air, fixations or a resonance chamber (Fig. 8.2).

Tension of the string by a force F results in a transversal displacement. The small shifted string piece Δx is pulled back into its equilibrium position by a force K_ξ, which is the difference of the force components in the vertical direction, acting tangentially at the ends of piece Δx (not considering elastic deformation), such that

$$K_\xi = F\sin(\alpha + \Delta\alpha) - F\sin\alpha, \tag{8.1}$$

where α is the angle formed between the tangent and the non-shifted string (along the x-axis).

For small shifts one has small α, thus

$$K_{\xi} = F(\alpha + \Delta\alpha) - F\alpha. \tag{8.2}$$

With $\Delta x, \Delta\alpha \rightarrow dx, d\alpha$ (infinitesimally small), one gets

$$\alpha = \frac{\partial\xi}{\partial x}$$

and

$$d\alpha = \frac{\partial^2\xi}{\partial x^2}dx. \tag{8.3}$$

The mass of piece dx is $\rho q dx$, where q is cross section and ρ is density of the string. The equation of motion is then

$$\rho q dx \frac{\partial^2\xi}{\partial t^2} = F\frac{\partial^2\xi}{\partial x^2}dx, \tag{8.4}$$

or

$$\frac{\partial^2\xi}{\partial t^2} = \frac{F}{\rho q}\frac{\partial^2\xi}{\partial x^2}. \tag{8.5}$$

with the general solution

$$\xi = f_1(x - c_{st}t) + f_2(x + c_{st}t),$$

where f_1 represents a wave traveling to the right, f_2 a wave traveling to the left, both with the propagation velocity on the string (c_{st})

$$c_{st} = \sqrt{\frac{F}{\rho q}}, \tag{8.6}$$

and consequently the frequency

$$f = \frac{c_{st}}{\lambda}, \tag{8.7}$$

where λ is the wavelength.

With length L of the string a basic frequency (f_b)

$$f_b = \frac{c_{st}}{2L} \tag{8.8}$$

is generated with vibrational nodes located at the end points of the string ($\lambda = 2L$).

For playing and tuning of string instruments, practical rules are:

- reducing length L increases frequency f: $(L/2 \to 2f)$
- the higher F, the higher f: $(4F \to 2f)$
- a thinner string causes increase of f: $(q/4 \to 2f)$

Precise experiments on these rules were already done by Vincenzo Galilei (around 1585) under assistance of his son Galileo.

8.1.3 Playing Techniques

There are several methods to make a string vibrate: plucking, striking and bowing (Fig. 8.3). For string keyboard instruments, see Sect. 10.2.1.

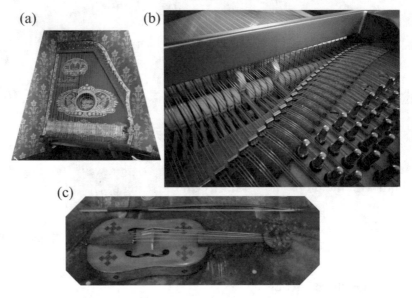

Fig. 8.3 a Zither—plucking (photo by KT in Volker Bley's gallery), **b** piano—striking (photo by KT), **c** vielle (in German Fidel)—bowing (public domain)

8.1.3.1 Plucking

Strings are "activated" with fingers, a plectrum or other mechanical devices. There are instruments without a fingerboard (harp, kithara, lyre) or with a fingerboard (guitar, lute, mandolin, banjo, zither (Fig. 8.3a), balalaika, ukulele, sitar, cembalo).

If a string is plucked at its center, the resulting vibration will consist of the fundamental plus the odd-numbered harmonics. All these modes have different frequencies of vibration, thus the shape of the string changes rapidly after plucking. The resolution of the string motion into two pulses that propagate in opposite directions on the string is illustrated in Fig. 8.4.

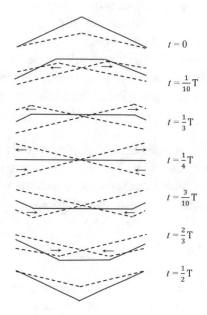

Fig. 8.4 Motion of a string plucked at its midpoint through one half cycle. T: period (Figure: taken from the book of Fletcher and Rossing mentioned in the Preface, Reprint permission from Springer)

8.1.3.2 Striking

Another method for sound production is to strike a string by small (felt) hammers, for instance, which is being used for the hammered dulcimer, but more importantly for the piano (Fig. 8.3b) or the clavichord.

Here we start with zero initial displacement, but a specified initial velocity. Important parameters are the mass and quality of the hammer. Its motion would be eventually stopped, and reflected impulses interact in a complicated way. Finally, the hammer is thrown clear of the string, letting it vibrate freely in its normal modes.

8.1.3.3 Bowing

This is a method used for many string instruments, including the violin family and the older viol family, as well as the medieval vielle (fidel) (Fig. 8.3c), the Chinese erhu, the Swedish nyckelharpa and others, where a bow is used for sound generation. This bow consists of a stick with a ribbon of parallel horse tail hairs stretched between its ends.

Figure 8.5 shows on the left schematically the string motion during bowing (from the tip to the frog—up bow) and on the right the displacements of the bow x versus time t. When the bow moves to one direction, a bend moves around, as shown by arrows in Fig. 8.5, left. This bend races around a curved envelope with a maximum amplitude that depends on the bow speed and the bowing position. The temporal displacement of the string has a saw-tooth like shape. The effect of moving a bow across the string is discussed in detail in Sect. 8.1.6.4.

The characteristic long, sustained, and singing sound produced by the violin family is due to the drawing of the bow against their strings. This sustaining of musical sound with a bow is comparable to a singer using breath to sustain sounds and sing long, smooth, or legato melodies.

8.1.3.4 Other Methods

The Aeolian harp is an unusual method of sound production: the strings are excited to vibrate due to the movement of the air. In a similar way, the Goura of South Africa is a musical bow which is not plucked, not bowed or struck, but blown with the mouth.

8.1.4 Bowed String Instruments

The great grandfather of string instruments was the bow. There are cave paintings from the stone ages showing scenes where music must have been made by using shooting bows. Before using this device for making a string vibrate, one started by plucking strings which led, for instance to the development of the harp. In the Middle Ages, around 900 years ago, some string instruments were

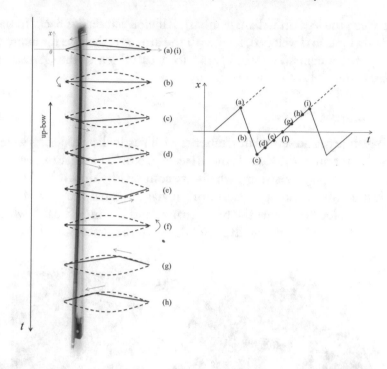

Fig. 8.5 Left: string motion during bowing, showing the shape of the string at 8 successive times during one cycle. Arrows indicate the direction of the bend racing. Right: a plot of the displacement x versus time. Letters correspond to the time axis illustrated in the left figure. For details see text (Figure: modified from the book of Fletcher and Rossing mentioned in the Preface, Reprint permission from Springer)

already played with help of a bow, e.g. the vielle. From these early examples several classes of string instruments evolved, two important of which are the viols and the violins.

Viola da Gamba

Around the beginning of the 16th century a family of string instruments came into use under the name "viola da gamba" (also called viol family: example in Fig. 8.6a). They may have originated in Spain with the popular vihuelas (name related to the denotation viola), which are plucking instruments considered as predecessors of the guitar.

There are various sizes for this instrument: treble, alto, tenor, bass, 2 sizes of double bass (violone).

The "gambas" have 5–7 strings and a fingerboard with frets at half-tone intervals. Strings are separated by a fourth. They are played vertically and are held between the legs (in Italian: gambas). All viols have a flat back. In parallel the viola da braccio developed, held with the arm as is the case for many modern string instruments. The word "braccio" (= arm) led to the German name "Bratsche", the viola of the violin family.

Viola d'Amore

In the Baroque period several instruments with particular properties were constructed, for instance, the viola d'amore (see Fig. 8.6b). This instrument usually has six or seven playing strings, which are sounded by drawing a bow across them, just as with a violin. In addition, it has an equal number of "sympathetic" strings located below the main strings and the fingerboard, which are not played directly but vibrate in sympathy with the notes played.

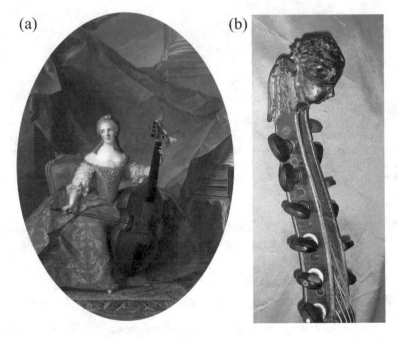

(a)   (b)

Fig. 8.6 **a** Viola da gamba. Princess Henriette de France as a young gamba player. The historical portrait was painted by Jean-Marc Nattier in 1754, Collection of Palace of Versailles (public domain), **b** head of a viola d'amore, early 18th century instrument, featuring blindfolded love (CC BY-SA 3.0 by Aviad2001)

This viola shares many features of the viol family. An intricately carved head at the top of the pegbox is characteristic on both viols and viola d'amores. Sound holes in the corpus are commonly in the shape of a sword known as "The Flaming Sword of Islam". It is unfretted and about the same size as the modern viola.

The first known mention of the name viol d'amore appeared in John Evelyn's diary (1679) [1]: "for its swetenesse & novelty the Viol d'Amore of 5 wyre-strings, …play'd on Lyra way by a German, than which I never heard a sweeter instrument or more surprizing … ". Leopold Mozart said that the instrument sounded "especially charming in the stillness of the evening."

After the late 17th century, the instrument fell from use, as the volume and power of the violin family became preferred over the delicacy and sweetness of the viol family. However, there has been renewed interest in the viola d'amore in the last century. The viola players Henri Casadesus and Paul Hindemith both played the viola d'amore in the early 20th century. Leoš Janáček originally planned to use it in his second string quartet, "Intimate Letters". The use of the instrument was symbolic of the nature of his relationship with Kamila Stösslová. Sergei Prokofiev's ballet Romeo and Juliet features a viola d'amore, as well.

8.1.5 The Modern Violin Family

A continuous development of string instruments followed towards the modern forms found nowadays in the violin family. Members of this family are: the violin, the viola, the cello, and the double bass. All of them have 4 strings tuned in perfect fifths, except for the double bass, where the tuning intervals are fourths (also, sometimes an additional string is added, going down to C1): violin: G3-D4-A4-E5; viola: C3-G3-D4-A4; cello: C2-G2-D3-A3; double bass: E1-A1-D2-G2. There are no frets on the fingerboards. The corpus of the instruments has approximate size of: violin—35.6 cm, viola—42 cm, cello—75 cm, double bass—105 cm.

Violin and viola are shown in Fig. 6.5 together with smaller versions (nominally 1/8, 1/4, 1/2, 3/4) of the violin for younger students. The cello has a shape which resembles closely that of the violin with bulging shoulders, whereas the double bass has mostly kept the shape of the bass viola da gamba with its typically sloped shoulders (see Fig. 8.7).

(a)   (b)

Fig. 8.7 a Cello (CC BY 3.0 by Georg Feitscher), **b** double bass (CC BY-SA 3.0 by An-drewKepert)

8.1.6 The Violin

8.1.6.1 History

With the use of the bow in the Middle Ages, new impulses arose in the Renaissance period. Based on predecessors like the rebec and the vielle, a first mention of the "violin" was made in 1523 with its name being a diminuitive of viola, coming from Italian. Its modern form developed in Northern Italy, due to the influence of several famous schools of violin makers (luthiers). Among these are the well-known names of Amati, Guarneri, Ruggeri, Montagna or Stradivari from the Cremona school, and others like the still famous Stainer from Tyrol, who dominated the scene for 150 years. Craftman's skills and artistic intuition entered into a unique symbiosis. In fact, especially the constructions of Stradivari (1648–1737) having a particularly beautiful and strong sound became very influential on the building of violins and pertain until today as a successful model. Violins from this "Golden Age" of violin making are nowadays of substantial value and the most sought-after instruments by collectors and performers (who may pay amounts in the million-dollar range). One of their secrets lies in the tonal quality, the large age and the specific varnish used.

Pioneering scientific contributions were made by Félix Savart about the harmonics elicited by the bowed string; by Ernst Chladni about the patterns of movement (see *Intermezzo* in Sect. 8.2.1); by Hermann von Helmholtz about the vibrations of a string; by Sir Chandrasekhara V. Raman on various aspects of the acoustics of violins, including a study of "wolf tones" (when the original note matches the resonant frequency of the body); and many others.

In the 19th century the length and tension of strings were augmented in order to obtain a stronger (but also harder) sound suitable for concert halls. Otherwise, there were no substantial changes of shape and structure of the violin until present, but it should be mentioned that industrial violin production has been introduced (e.g., by M. Suzuki) and more experimental versions were being built, for instance, violins with trapezoidal shape or made of synthetic substances.

8.1.6.2 Violin Construction

In Fig. 8.8 we present a sketch for a violin with indications of some relevant parts.

Body

The corpus of a violin is made of two arched plates fastened to ribs, the sides of the "box". The top is made of spruce, which has a very high elasticity in relation to its low weight. It has two sound holes (f-holes) which allow the box to breathe as it vibrates. The bottom is mostly carved from maple, often with a

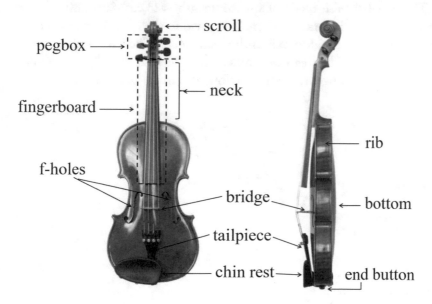

Fig. 8.8 Names of the parts of a violin (CC0, photo by Just plain Bill)

matching striped figure, called "flame". Four strings are anchored to the lower bout of the instrument.

Neck

The neck is made from the same material as the back and the ribs. It carries the fingerboard, typically made of ebony, preferred because of its hardness and resistance to wear. At the upper end of the instrument we find the pegbox with one peg for each string, also made of ebony. The scroll at the end of the pegbox provides essential mass to tune the fundamental body resonance.

Bridge

This precisely cut piece of maple forms the lower anchor point of the vibrating strings. It transmits the vibrations to the body. Its top curve holds the strings at the proper distance from the fingerboard. Its mass distribution has a prominent effect on the sound.

Sound Post and Bass Bar

Two elements with important acoustic function are placed inside the body of the violin (Fig. 8.9). The sound post is a cylindrical stick made of spruce, fitted between top and bottom of the body, just to the tailward side of the bridge foot. It supports the top under string pressure, transmits vibrations and equilibrates the direction of vibration between upper and lower part.

The bass bar (also made of spruce, length ≈23 cm) is running under the opposite side of the bridge, where it is glued to the cover in parallel direction with respect to the middle axis of the instrument. The bar stabilizes the statics of the cover plate. Its shape and mass affect the tone at the lower frequencies in a way that improves the well-distributed warmth of the violin.

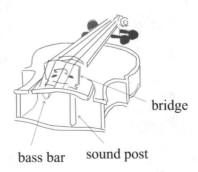

bridge

bass bar   sound post

Fig. 8.9 Cross section of a violin

Strings

Among the 4 strings the uppermost one (E5) is made of steel. The others were traditionally based on gut, which is sensible with a warm tone, but has a relatively long playing-in time. Modern strings have commonly a stranded synthetic core wound with various metals of several layers, like aluminium, silver, gold or tungsten, thus enhancing warmth and brightness. The artist may choose among many options and then decide which strings correspond best to his expectations. In fact, material properties of strings and their tension, thickness and length have substantial influence on the sound of violin.

Intermezzo

The pegs are an indispensable part of certain stringed instruments. A properly working peg will turn easily and hold reliably, that is, it will neither stick nor slip. Often one uses a substance called "peg dope" to coat the bearing surfaces.

However, tuning remains a task where some force must be exerted by the hands of the musician. For the young Yehudi Menuhin - at the age of 11 years -, during a rehearsal of the Beethoven violin concerto with the New York Symphony Orchestra at Carnegie Hall, it turned out that he had to ask the concertmaster to tune his violin, because his hands did not have the necessary power. Nevertheless, after the concert Yehudi had entered the world stage of artists (and he had to go to bed right afterwards).

When he was 14 years old, he played a violin concert of Mozart under the conductor Bruno Walter in Berlin (the photograph below: Bundesarchiv, Bild 102-12786 / CC-BY-SA 3.0). We are confident that he could then tune the violin by himself.

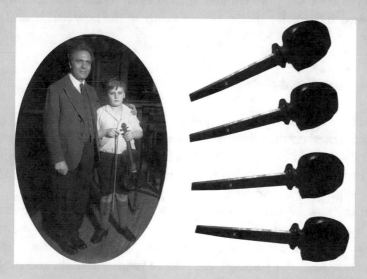

8.1.6.3 The Bow

The bow is just a long piece of wood or a stick with a hank of horsehair (preferentially from the tail of a white horse) strung between the tip and frog at opposite ends, as shown in Fig. 8.10. The tip of the bow is a part of the stick and accommodates the upper end of the hair ribbon. For the frog, which holds and adjusts the near end of the horsehair, ebony is most often used. The frog with its characteristic shape contains a screw to adjust the tension of the bow's hairs.

The stick was traditionally made of pernambuco or brazilwood. The length of a violin bow (with screw) is 75 cm, and its weight is 58–62 g. The bows for the viola and the cello are slightly larger. For the double bass of the German form[1] it has a total length of 68.5 cm and weighs ≈125 g.

In earlier times (especially in the Middle Ages and the Renaissance era) the stick had a concave shape (see the round bow of the vielle in Fig. 8.3c). This form enables players to control the tension of the bow hair in order to play not only one but also two, three or four strings simultaneously. As seen in Fig. 8.10 modern bows consist of a tensioned straight stick with a slightly convex curvature.

The hair must be occasionally rubbed with rosin (colophony), so it will grip the strings. The action of the bow depends almost entirely upon the application of this material and upon its frictional properties. Violin rosin is a natural gum obtained from conifers such as larch which produce turpentine. When powdered, small particles adhere firmly to the bow hair because of ion exchange.

8.1.6.4 The Bowed String

Pulling a bow across a string is a process involving some interesting physics. Helmholtz observed that the displacement of the string forms a triangular shape, as illustrated in Fig. 8.11 [2]. When a bow is moved on the string, for a certain time, the bow hair sticks on the string because of the friction, and then suddenly slips: after a short time it sticks again, and then slips.

When the bow hair and the string stick and move together, their relative velocity is zero, and there is a maximum tangential force between the bow hair and the string. Such a friction is called "static friction". On the other hand, when the bow hair slips (sliding), the relative velocity is high, and the friction is called "dynamic friction".

---

[1]The form of the frog in French bows for a double bass is different from that for the German one.

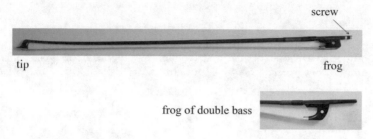

Fig. 8.10 Top: Bow for a violin (CC BY-SA 3.0 by Arent), bottom: frog of a German double bass bow (CC BY-SA 3.0 by Bottesini)

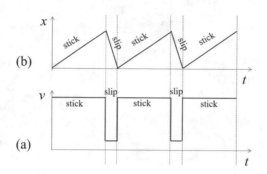

Fig. 8.11 a Velocity of a bow on the string, b displacement of the string

When a stick-slip process is repeated during bowing, the displacement of the string assumes approximately a saw-tooth shape. Such a saw-tooth function can be transformed to a frequency spectrum via a Fourier series, as described in Sect. 6.3.1. This function generates a typical sound spectrum of a violin, which reaches very high frequencies.

The sound created by bowing depends on the bow speed, the pressure of the bow on the string, as well as on the place of the string where bowed. According to Cremer [3], both the maximum displacement of the string and the peak transverse force at the bridge are proportional to the ratio between the bow speed and the distance from the bridge to the bowing point.

The range of the bow force required for the sound to be audible or tolerable, depends on the bowing position. The closer the bow position to the bridge, the stronger the force, and the harder the created sound. On the other hand, sound created from a bow position far from the bridge is softer.

Coffee Break

There is a pig in your violin! There is a slightly bigger pig in your viola! There is a still bigger pig in your cello! There is a giant pig in your double bass! Some of them are friendly, others are angry, thoughtful, or lonely,.... Do you know that?

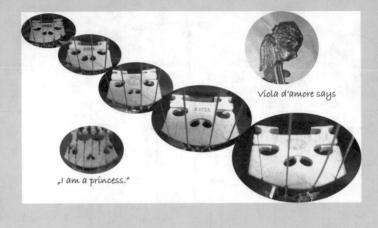

8.1.7 Sound and Acoustics of the Violin Family

Luthiers have made violins of exceptionally beautiful sound from purely practical experience. Here is no chance for physics to explain which ways they chose. Of course, since the 18th century scientists tried to understand their secrets, but many questions have not yet been solved even in present time. And luthiers have followed what experience suggests them to do.

We will summarize the acoustic characteristics of the resonance body of a violin. The bottom and cover have different vibration characteristics, which is essential for the sound emanating from the whole body.

The French physicist and medical doctor Félix Savart asked how large the distance between the cover and bottom should be, and he applied the Chladni method [4] (see *Intermezzo* in Sect. 8.2.1). He found that the resonance of the cover is at the pitch between C♯ and D, and that of the bottom between D and D♯. In a good violin such a difference in the resonance of the cover and bottom is found. When there is no difference, the sound becomes "rough". The thickness of the plates, vibration forms, the volume of the resonance body, as well as flexibility and varnish are interrelated, resulting in the complex timbre.

In addition to Chladni's method interference patterns can supply more details such as the vibration amplitude with a resolution of μm. There is a

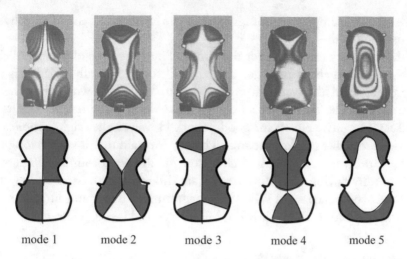

mode 1   mode 2   mode 3   mode 4   mode 5

Fig. 8.12 Bottom: simplified vibration pattern of modes 1 to 5. The black lines are the nodal lines, white and blue areas show where the vibration would be opposite in phase, top: corresponding interference patterns from [5], reprint permission from Springer Nature

clear difference of interference structures and modes of the top and bottom covers, according to the different resonance frequencies. In Fig. 8.12 a mode sequence of the bottom cover is depicted.

Luthiers tap the violin and listen to the excitation of a special mode. Although all modes contribute more or less to the sound, the modes 1, 2 and 5 are especially important. (Note that these modes are comparable, but not the same as the modes for the oscillation of an ideal circular membrane as presented in Sect. 8.2.1.1.)

How to assemble all parts is so complicated that no method can follow it with acoustical means. There are many parameters to be considered, and they should be checked at different construction stages. Following four features should be respected for obtaining high quality:

- The frequency of mode 1 should be about one octave lower than that of mode 2.
- Good instruments exhibit mode 5 with high amplitude. Frequency difference between the cover and bottom is about 1 tone.
- If the mode 2 deviates in frequency with less than 1.5%, the sound is softer.
- It is good, when the frequencies of mode 2 and mode 5 are different by about one octave.

These are purely empirical findings, which help luthiers. But there is no physical explanation for these points.

Varnish is very important for the sound. Since the time of Stradivari there is a big question about this, not solved until today. Actually, varnish is used for protection of the wood. It covers the surface of the wood, but there is no varnish inside. Since varnish changes the wood to be less elastic, its use could have negative effects for good sound. However, the "right" application of varnish somehow makes the sound better. It is said that it takes a long time till the varnish obtains the final character for the good sound. Temperature and humidity during varnishing/drying are also important, because they affect the water/wood exchange. When the environment is hot and humid, sound becomes monotone. When it is hot and dry, sound is rough and hard.

Vibrato—a Further Important Aspect

Vibrato is essential for playing not only violins but also violas and other string instruments. It is a pulsed effect in pitch, serving an additional expression, and produced by vibration of a finger, hand or underarm, depending on the musical taste. Vibrato causes a pitch variation of up to ±50 Cent together with a feeling of tempo change.

In singing, vibrato is caused by nerve-related interactions. String instruments and wind instruments imitate this function, because vibrato is a stylistic element which provides warmth and quality in sound. Furthermore, at higher pitches the direction of the sound propagation is changed, so that the sound becomes silvery and subsequently more transparent against the orchestra.

8.1.8 Sound of Viola, Cello and Double Bass

Violas, cellos and double basses belong to the violin family and they are, in many respects, similar to each other.

8.1.8.1 Viola

A viola is a bigger violin, but its proportion is slightly different from a violin. It sounds lower, as mentioned in Sect. 8.1.5. Each string is a fifth lower compared to those of a violin, while the ratio of the corpus sizes between a viola and a violin is 42:36 (=7:6). Conceptually the ratio should be 54:36 (=3:2), but then the instrument would be too long for a human arm. Subsequently, the sound of a viola has a character different from that of a violin. The viola does not have the "e" string (the highest string of the violin) which supplies a bright power, but the "c" string (the lowest of a viola) which gives a slightly harsh taste.

The sound of a viola is, therefore, full, soft, dark, and in the higher positions, melancholic, husky having a somewhat nasal timbre.

8.1.8.2   Cello

All strings of a cello are an octave lower than those of a viola. It means that theoretically the corpus length of a cello should be three times longer than that of a violin. However, the real size is only twice longer. Instead, the total length is much larger compared to that of a violin, which creates strong higher partials, resulting in warm tone color.

A cello has a wide range of pitch—almost 5 octaves, which is, however, not so clearly radiated. Sound characteristics of each string is:

- c string: emphasizing bass tones, dark
- g string: brighter and softer
- d string: solo-player's strength, rich in overtones
- a string: bright, radiating, with tone color often similar to viola

The fundamental is rather weak compared to the harmonics (15 dB lower than the strongest harmonic). Sounds are full of variety due to the different positions, suggesting the different formants.

8.1.8.3   Double Bass

The form of a double bass is a mixture of violin and gamba and, therefore, it may not belong to the violin family in a narrow sense. Moreover, the tuning (E1-A1-D2-G2) is in fourths (not fifths). Instead of tuning pegs a double bass has a mechanism of metal scrolls. The bowing technique is also different. As shown in Fig. 8.10, a frog of the German double bass bow is quite different from that of a violin.

The length of the string from the bridge to the upper end of the neck (scale length) is ≈105 cm for a 3/4 bass and ≈107 cm for a full-size 4/4 bass. Bass players can choose one of them according to their body size.

8.2 Percussion Instruments

From the point of classification percussion instruments can be divided into membranophones and idiophones (see Chap. 6). Instruments belonging to the membranophones are various drums, timpani and tambourines. Most of other percussion instruments are idiophones. They are, for examples, bells, cymbals and xylophones. In this section we introduce timpani and drums for membranophones (used mainly in orchestras) and bells for idiophones.

8.2.1 Timpani and Drums

Upon hitting a drum or timpani, some kind of sound (for most of cases perceived as noise[2]) is created due to the vibration of the membrane.

Timpani can be tuned. Figure 8.13 shows a timpani set in an orchestra with their pitches depending on size.

A drum kit (see Fig. 8.14) is commonly not used in an orchestra, but in a rock band. In principle, the drums can not be tuned and just create rhythmical noise.

As mentioned in Sect. 8.1 the vibration of a string is a 1-dimensional phenomenon. On a vibrating string there are points where no oscillation occurs. These points are called nodal points. The sound is determined by the harmonic spectra (multiples of the fundamental). On the other hand, the vibration of a membrane is 2-dimensional, and therefore, there are some lines where no oscillation is taking place. These are called nodal lines. The created sound is more complex and in general not harmonic.

In order to understand such sound (or noise) we will explain below, how a membrane vibrates, when it is set and strained homogeneously on a circular frame like an ideal drum head.

8.2.1.1 Vibrations of a Circular Membrane

The vibration of a circular membrane can be described by the following wave equation of the oscillation amplitude $u = u(x, y, t)$:

$$\frac{\partial^2 u}{\partial t^2} = c^2 \left(\frac{\partial^2 u}{\partial x^2} + \frac{\partial^2 u}{\partial y^2} \right), \tag{8.9}$$

---

[2]Noise is a sound with a continuous frequency spectrum comprising many kinds of sounds. See Chap. 2.

Fig. 8.13 Left: timpani set in an orchestra (CC BY-SA 3.0 by Flamurai), right: pitch ranges for various sizes of standard instruments and extended instruments (small notes at the right end of each bar) (public domain by Number Googol)

Fig. 8.14 Example of a standard drum kit: a bass drum, a snare drum, a floor tom, a pair of tom-tom drums, a hi-hat, a crash cymbal and a ride cymbal with a stool (called throne) and sticks (CC BY-SA 2.0 by Mark dolby)

where x and y are spatial coordinates in the plane of the membrane, t is time and c is the propagation velocity of the wave. If the membrane is fixed around a circular frame as is the case for timpani and drums, the vibration amplitude is zero on the frame at any time. It means that $u = 0$ at the boundary (this is the boundary condition).

Equation 8.9 is presented in a Cartesian coordinate system. However, for a circular membrane, it is advantageous to use a polar coordinate system with r (radius) and θ (angle) with the center of the membrane at $r = 0$. Then the wave equation in polar coordinates is written as:

$$\frac{\partial^2 u}{\partial t^2} = c^2 \left(\frac{\partial^2 u}{\partial r^2} + \frac{1}{r}\frac{\partial u}{\partial r} + \frac{1}{r^2}\frac{\partial^2 u}{\partial \theta^2} \right) \tag{8.10}$$

with $0 \leq r < a$ and $0 \leq \theta < 2\pi$. The boundary condition is $u = 0$ for $r = a$.

A solution of this wave equation is the membrane oscillation:

$$u(r, \theta, t) = A \cdot J_m(kr) \cdot cos(m\theta)sin(\omega t), \tag{8.11}$$

where $k = 2\pi/\lambda$ (wave vector) and $\omega = 2\pi f$ (circular frequency).

Here J_m is the Bessel function of the mth order. As shown in Fig. 8.15, with increasing kr, the Bessel functions of various orders (from 0 to m) become close to the sine function.

From the boundary condition, $u(a, \theta, t) = 0$. Now let us look at the red curve in Fig. 8.15. This is a plot of the Bessel function of the 0th order J_0 ($m = 0$). One sees that there are many (an infinitive number) of ka's at which $J_0 = 0$. These k's are denoted as k_{0n}, numbered with n (n = 1, 2, ...). The same is true for other Bessel functions (m = 1, 2, ...): there are ka's at which J_m is 0. These ka's are numbered also with n (n = 1, 2, ...). We can describe such ka's as,

$$k_{mn}a \ (= n-\text{th root of the Bessel function of the m}-\text{th order}),$$

where

$$J_m(k_{mn}a) = 0.$$

Each combination of m and n, (mn), is called "mode", for each of which standing waves are built. For example, the mode (01) is the case of m = 0 and n = 1. Patterns of such standing waves are called sound figures, nodal patterns or **Chladni figures**.

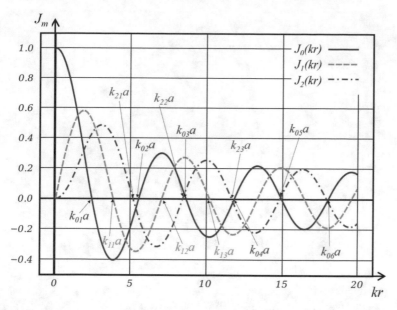

Fig. 8.15 Bessel functions as a function of kr (modified from a picture of public domain by Inductiveload)

Table 8.1 lists some root values of the Bessel functions.

Table 8.1 Roots of Bessel functions $k_{mn}a$ (6 digits)

m\n	n=1	n=2	n=3	n=4	n=5	...
m=0	2.40483	5.52008	8.65373	11.7915	14.9309	...
m=1	3.83171	7.01559	10.1734	13.3237	16.4706	...
m=2	5.13562	8.41724	11.6198	14.7960	17.9598	...
m=3	6.38016	9.76102	13.0152	16.2235	19.4094	...
m=4	7.58834	11.0647	14.3725	17.6160	20.8269	...
...	...	...	...	...	...	...

Intermezzo

Ernst Chladni (1756–1827) was a German physicist and musician. He developed a technique to show various patterns which are caused by different modes of vibration on a rigid surface: drawing a violin bow over a piece of metal whose surface was lightly covered with sand, as shown below. He published his experimental results in "Entdeckungen über die Theorie des Klanges (Discovery on the sound theory)" in 1787 [6]. These patterns have been called "Chladni figures" ever since.

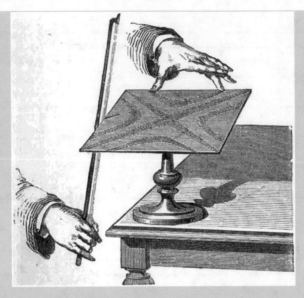

Source: William Henry Stone (1879) Elementary Lessons on Sound, Macmillan and Co., London, p. 26, Fig. 12 (public domain)

Although these are called "Chladni figures", he was not the first person to find such patterns. Galileo had observed vibrational patterns on a brass plate in 1638 [7]. A little later, in 1680, Robert Hooke observed such nodal patterns associated with the vibrations of glass plates. Hooke had also run a violin bow along the edge of a plate covered with flour [8].

Chladni demonstrated the vibration patterns at the Paris Academy in 1808. Napoleon was one in the audience of leading scientists. Napoleon set a prize for the best mathematical explanation. Sophie Germain answered with the correct approach, but her answer was rejected, because it was thought to be a flaw.

8.2.1.2   Modes and Nodal Lines

Now we will find the nodal lines for each mode.

Case a: m = 0

From Eq. 8.11, when m = 0,

$$u(r, \theta, t) = A \cdot J_0(kr) \cdot cos(0)sin(\omega t) = A \cdot J_0(kr) \cdot sin(\omega t). \quad (8.12)$$

Therefore, for $r = a$ nodal lines $u(a, \theta, t)$ should fulfill

$$J_0(k_{0n}a) = 0. \quad (8.13)$$

Since Eq. 8.13 is independent of θ, the solution is a circle. And from Table 8.1: $k_{0n}a \approx 2.405, 5.520, 8.654, 11.793, ...$ for n = 1, 2, 3, 4, ..., respectively.

When n = 1, the nodal line is a circle with radius a. This frame ring corresponds to $k_{01}a$ (≈ 2.405). It is the mode (01) and is called the fundamental mode.

When n = 2, the frame ring corresponds to $k_{02}a$ (≈ 5.520). In this case, there is an additional smaller ring, the radius of which is $(k_{01}/k_{02})a \approx 0.44a$.

With higher n's, the frame ring corresponds to $k_{0n}a$, and there are (n-1) smaller rings within the circle of radius a. Their radii are $(k_{01}/k_{0n})a$, $(k_{02}/k_{0n})a$, ... , $(k_{(0(n-1))}/k_{0n})a$. These ring modes are shown in Fig. 8.16 as modes (01), (02), and (03).

Case b: m = 1

When m = 1, Eq. 8.11 writes

$$u(r, \theta, t) = A \cdot J_1(kr) \cdot \cos(\theta)\sin(\omega t). \qquad (8.14)$$

Therefore, nodal lines should be either

$$J_1(k_{1n}a) = 0 \qquad (8.15)$$

or

$$\cos(\theta) = 0 \qquad (8.16)$$

Equation 8.15 is similar to Eq. 8.13 and presents rings. On the other hand, Eq. 8.16 is fulfilled only when

$$\theta = \frac{1}{2}\pi \quad \text{or} \quad \frac{3}{2}\pi.$$

This is a straight line through the membrane center. As shown in Fig. 8.16, the mode (11) possesses one straight line through the center of the membrane and one ring (frame). The modes in which nodal lines are drawn straight through the membrane center are called radial modes. The mode (12) has one straight line through the center of the membrane and 2 rings.

Case c: m = 2

For m = 2, Eq. 8.11 becomes

$$u(r, \theta, t) = A \cdot J_2(kr) \cdot \cos(2\theta)\sin(\omega t). \qquad (8.17)$$

Therefore, nodal lines should be either

$$J_2(k_{2n}a) = 0 \tag{8.18}$$

or

$$cos(2\theta) = 0 \tag{8.19}$$

Equation 8.18 is similar to Eq. 8.13, and presents rings. Equation 8.19 is fulfilled only when

$$\theta = \frac{1}{4}\pi, \quad \frac{3}{4}\pi, \quad \frac{5}{4}\pi, \quad \text{or} \quad \frac{7}{4}\pi.$$

This defines 2 straight lines through the membrane center which are orthogonal to each other. As shown in Fig. 8.16, the mode (21) possesses 2 straight lines through the center of the membrane and one ring (frame), the mode (22) 2 straight lines and 2 rings.

Case d: m = 3, 4, …

When m increases, m equi-angular nodal lines through the membrane center appear. Some examples (31), (32), (41), (51) and (61) are illustrated in Fig. 8.16.

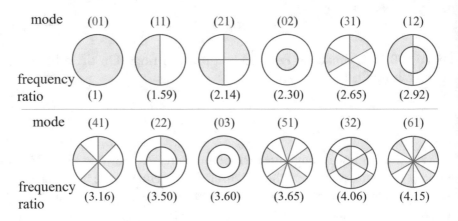

Fig. 8.16 Nodal patterns of various modes and the corresponding frequencies relative to the fundamental one. Red circles are nodal circles and blue lines are nodal diameters. Except (01) there are two phases in their oscillations, the areas of which are marked with light green and white. (01), (02) and (03) are ring modes, while (11), (21), (31), (41), (51) and (61) are radial modes

8.2.1.3 Mode and Frequency

The modal frequencies of the circular membrane f_{mn} are

$$f_{mn} = \frac{\omega_{mn}}{2\pi} = \frac{ck_{mn}}{2\pi}.$$

The speed of propagation of transverse waves on a circular membrane c is

$$c = \sqrt{\frac{T_l}{\sigma}}$$

where T_l (N/m) is the surface tension per unit length of the membrane and σ (kg/m^2) is the areal mass density of the membrane.

Therefore,

$$f_{mn} = \frac{1}{2\pi}\sqrt{\frac{T_l}{\sigma}}k_{mn} = Fk_{mn}, \tag{8.20}$$

where F is a constant determined by the surface tension and the areal mass density of the membrane. Thus, the ratio of the frequency for the mode (mn) to the fundamental is k_{mn}/k_{01}. Corresponding ratios are calculated by using Table 8.1, and they are listed below each pattern in Fig. 8.16.

Different from the case of the string the vibration of a membrane is not harmonic, because the ratios of frequencies for each mode are not in an integral relationship: 1 to 1.59, 2.14, 2.30, ... This means that membrane vibrations cannot create sound with a certain pitch. They are "noise".

But if we think about the sound of timpani, there is a contradiction. We do hear sounds with certain pitches.

8.2.1.4 Timpani

If timpani would create only noise, they could not play an important role in orchestras. Timpani can be tuned over a range of more than an octave by 6–8 screws on the frame. Modern timpani have additionally a clamping device which is operated by a pedal (see Fig. 8.13).

In praxis it is known that timpani create different sound/noise depending on the place where they are struck. If the center is struck, it sounds like cardboard boxes. This indicates that the mode (01) creates mainly noise. If a place close to the edge is struck, only very weak and thin noise is heard, suggesting that to strike the first nodal ring does not create sound.

In most of tutorials for learning timpani it is written that the best place to strike is at about 3/4 of the radius away from the center. If the right spot is hit, the tone is nice and full with a clear pitch. To strike such a place seems to cause membrane oscillations of a mode other than (01).

In 1945 Rayleigh concluded from his experiments that the main tone of the timpani is generated from the mode (11) and that the modes (21), (31) and (12) are its harmonics [9]. If the mode (11) is considered as the fundamental, frequencies of the modes (21), (31) and (12) are 1.5x, 1.88x and 2.0x that of the fundamental, respectively. Later Rossing and Kvistad obtained similar results by using modern equipment: Frequencies of modes (21), (31), (41) and (51) are approximately 1.5x, 2x, 2.5x and 3x that of the mode (11) [10], although theoretical values are 1.34x, 1.66x, 1.98x and 2.29x.

Several reasons exist for the discrepancy between the ideal membrane theory and praxis. These are effects of air loading, air enclosed by the kettle, bending stiffness and stiffness to shear. For a thin membrane, the effect of air loading is most dominant, lowering the modal frequency.

There are the piston approximation [11] and the Green function method [12] for calculating the air loading. In the piston approximation the effective air mass loading for a piston in an infinite baffle is considered. A more accurate determination of the effect can be made by applying the Green function technique. Expressions can be derived for the pressure inside (in the kettle) and outside the membrane. The results of the ratios of the frequencies for modes (21), (31), (41), (51) and (61) to the fundamental frequency (11) are 1.51, 1.99, 2.46, 2.93 and 3.38, respectively. These values agree quite well with the experimental results.

The volume of the kettle also affects the membrane modal frequencies. When the volume is reduced, the frequencies of axially symmetric modes (01). (02) and (03) are raised, while those of other modes are lowered.

One more factor which is important in praxis is the time evolution of the timpani spectra. Upon striking a timpani its sound spectrum shows an undefinable continuous curve consisting of frequencies of various modes. Many of these components decay quickly (within half a second). This is called an "impact noise". After ≈0.5 s only slowly decaying partials remain. The decay times of these partials are different from each other (7–10 s). When the impression of the "impact" is important, percussionists touch their hand on the membrane to damp vibrations at ≈0.7 s for low pitches and at ≈0.2 s for high pitches [13].

Also, in general, the modes (11), (21), (31) and (41) decay slowly compared to the modes (01), (02) and (03) [10].

8.2.1.5 Drums

Different from timpani, drums are generally unpitched percussion instruments. It means that they create "noise". However or therefore, they play important roles for rhythms and accents in music. There are many kinds of drums around the world from early times on: bass drums, snare drums, tom-toms, bongos (see Fig. 6.2 in Chap. 6), Indian drums, Japanese drums, as well as various drums in Africa, in Latin America, ...

Here we will explain two of them which are often used in European orchestras: the bass drum and the snare drum.

Both drums are cylindrical with two membranes at both ends. The diameter of the drum is much larger than its depth. Because of this form there is an interaction between the upper membrane (batter or beating head) and lower membrane (carry or resonating head), as illustrated in Fig. 8.17. Such an interaction can be explained by a simple two-mass oscillator for a bass drum, while for a snare drum the interaction is more complicated due to the attached snare.

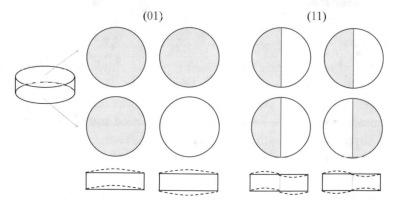

Fig. 8.17 Interaction of the beating and resonating heads for modes (01) and (11)

Bass Drums

The diameter of the concert bass drum is 80–100 cm.

Usually a soft mallet is used for avoiding the excitation of high frequencies. The normal impact location is close to the center, which causes ring modes of the membrane vibrations, creating a muffled tone with a frequency range of 100 Hz (mixture of inharmonic partials).

In most of the cases the beating head is tuned with a greater tension than the membrane on the opposite side (resonating head). But sometimes they are tuned the other way around or with equal tension.

One interesting phenomenon is a coupling effect between two heads, especially when both membranes are at the same tension. For example, two frequencies, 44 Hz and 104 Hz, are observed for the mode (01). When the beating head oscillates with a frequency f_1 in the mode (01), the resonating head oscillates with a different frequency f_2 in the same mode:

$$f_2 = \sqrt{f_1^2 + 2f_c^2},$$

where f_c is the coupling frequency which depends on the stiffness of the air volume and the membrane mass. In our case, with $f_1 = 44$ Hz and $f_2 = 104$ Hz, one has $f_c = 67$ Hz.

When two membranes vibrate with different modes in which frequencies are very close, beats (see Chap. 3, Sect. 3.1.3) are heard. This sounds like a breathing of the bass drum.

Snare Drums

When dealing with a snare drum, many people think of Maurice Ravel's "Boléro" (Chap. 14), where it plays a leading role. Since snare drums have a short transient time of about 7 ms, and also since there is little contribution of low frequency (the maximum of the radiated sound: 300–1000 Hz), they can supply rhythm with high precision [13]. Therefore, since Gioachino Rossini's opera "La gazza ladra" a snare drum has been used in orchestras as a rhythm instrument (Fig. 8.18).

The orchestral snare drum has a form self-similar to the bass drum, but smaller (diameter: ≈35 cm, depth: 13–20 cm). Moreover, it has strands of wire (snare strings) stretched across the resonating head. The snare strings collide with the resonating membrane, exciting additional vibrational modes of high frequencies. The manner of the collision depends on the mass and the tension of the snares. The snare tension is optimum, when the snare and membrane move in opposite directions with maximum speed at the moment of contact.

Fig. 8.18 Left: view of a snare drum with strings at the resonating membrane, right: enlarged view of the snare strings (photos by KT)

Then the impact is large and, therefore, one hears the characteristic "noise" of a snare.

Rossing et al. observed oscillations of the whole drum (not only the two membranes and snare, but also the cylinder body) by holographic interferograms, while a snare drum is played. The amplitude of the oscillation for the cylindrical body is about 1% of that for the membranes [14]. This could explain the very "nice" and complicated noise of the snare drums.

8.2.2 Bells

A bell can be tuned. Most of bells are made from metal, and once formed they stay as they are, as inflexible as statues or monuments. It means that bells should be already tuned when they are born. Their pitch depends on the size, shape, material, as well as thickness distribution.

Old Chinese people took such "inflexibility" of bells as an advantage for preserving tones. The famous bell set "Bianzhong of Marquis" was found during the excavation of the tomb of Marquis Yi. This set consists of 64 bells, which preserves 12 half tones of more than 5 octaves. The form of the Bianzong bells is unique: an oval form with round cuts at the bottom (Fig. 8.19c). In each bell there are two fundamental tones which are created by hitting two different places. For details of mechanism and structure see [15]. (See also Sect. 5.4.1 and Sect. 11.3 in Fig. 11.7.)

Before mentioning European bell tuning we just show some different shapes of bells in Fig. 8.19. In Europe most of the bells have a form with a flared opening like the Zygmunt Bell, carillon bells and hand bells (Fig. 8.19a, d), while bells used for the Buddhist temples in South-east Asia and East Asia have a straight opening as recognized in Fig. 8.19b. Small round bells are seen in many places from the early times. Such bells made with clay about 4500–3300 years ago were found in Japan. Later they were cast from metal. One example is the pre-Columbian bell made by North American natives (see Fig. 8.19f).

Many of such round bells are connected together on a stick, a ring or a piece of rope and used as a music instrument (a sleigh bell or jingle bell). A famous musical piece using sleigh bells is Deutscher Tanz (German dance) KV 605, Nr. 3—"Schlittenfahrt (Sleigh ride)" of Wolfgang Amadeus Mozart. His father Leopold composed "Musikalische Schlittenfahrt" and used a sleigh bell, as well. In Richard Eulenberg's "Petersburger Schlittenfahrt", too, a sleigh bell can be heard. For the sleigh ride, it seems that the sound of jingles is indispensable.

Fig. 8.19 Various bells **a** the Zygmunt (Sigismund) Bell (from 1520) in Kraków, Poland (CC BY 2.5 auhor unknown), **b** Japanese temple bell of the Ryoanji Temple, Kyoto (CC BY-SA 4.0 by Syohei Arai), **c** Chinese bell, Bianzhong about 800 BC (public domain), **d** carillon bells, Peter and Paul Cathedral, St. Petersburg (CC BY-SA 3.0 by RuED), **e** English hand bells (CC BY-SA 3.0 by Oosoom), **f** pre-Columbian bell made by North American natives (public domain, by Prof. Cyrus Thomas), and **g** a sleigh bell for children (photo by KT)

8.2.2.1 European-Style Bells

History

Figure 8.20 illustrates a modern bell (a) and its cross section (b). It is (or at least, tries to be) axially symmetric. A section of an the old bell is also shown in blue for comparison. A bell with the old shape is called a beehive bell. This form was common between the 8th and 12th centuries. In most of the cases the wall was thin: all parts of the section had more or less an equal thickness.

Since the beehive bells can not create harmonic tones, the shape of the bell has been modified through the years (for example, a sugar loaf shape, a Gothic shape, a Baroque shape and so on) in order to obtain "nice sounds". And finally, in the 1850s, the modern shape was established, which creates an optimum sound pattern. Being axially symmetric is important for avoiding beats. Also people knew empirically that the alloy of copper (80%) and tin (20%) provides the most pleasant tones.

Bells with Pleasant Sound

When a bell is struck by a clapper, it vibrates in a very complicated way. Similar to the vibration of a membrane the number of modes is infinite, not 2-dimensional but 3-dimensional, though. A few of these modes contribute to the audible sound.

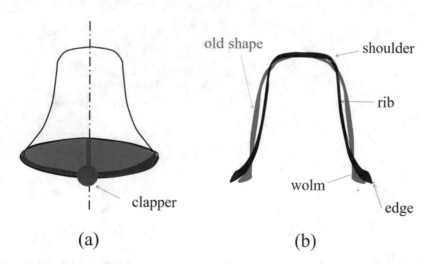

Fig. 8.20 **a** A modern European bell, **b** sectional view of a modern European bell (in black) and that of an old bell (in blue)

In between it has been known that the proper shape creates the sound, so that the bell's strong harmonics are in harmony with each other as well as with the strike note.

The main harmonics of a well-tuned bell are:

- hum (an octave below the prime)
- prime (also called strike tone, tap note or named note)
- tierce (a minor third above named note)
- quint (a fifth above named note)
- nominal (an octave above named note)

A relationship between these harmonics and modes with nodal lines is illustrated in Fig. 8.21. Modes are described in parentheses with numbers m and n as (mn). The "m" corresponds to the number of the meridian nodal lines, while "n" to the circular nodes. The number of the meridian pair increases with increasing m. When "n = 1", the position of the circular node can be either close to the bell opening or to the bell tip. The pitch created in the former case is higher than in the latter case. Therefore, it is marked as $n=1_{\#}$ (for lower circle node) or $n=1$ (for higher circle node). The suffix $\#$ is also used for the higher m. Here we assume that the material of the bell is hard enough so that the length of the ring does not change (inextensional).

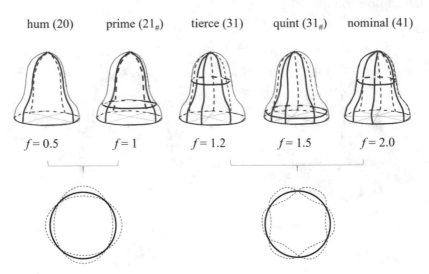

hum (20)   prime ($21_{\#}$)   tierce (31)   quint ($31_{\#}$)   nominal (41)

$f = 0.5$ $f = 1$ $f = 1.2$ $f = 1.5$ $f = 2.0$

Fig. 8.21 Top: the five strongest modes of a bell. Modes are described in parentheses (mn), bottom: schematic views of vibration at the bell opening

The strike tone (prime) is the dominant note perceived immediately by the human ear. This tone is, however, not so strong as the other four audible tones. Only when a bell is properly struck, this note prominently attracts the attention of the ear. What we actually hear is a mixture of harmonics which are generated at a strike. They are notes of an octave higher (nominal) and lower (hum), minor third (tierce) and perfect fifth (quint). This is the reason why the sound of bells is in minor.

The mode ($41_\sharp$) is the next higher mode and this is a major third. Theoretically it is possible to strengthen this mode. There have been attempts to produce a "major" bell in the 1980s. Subsequently some major bells with specific forms were made. However, most of the people prefer to hear the traditional minor sound.

If you run your moistened finger around the rim of a wine glass, you hear sound of $m=2$ mode. If you like to know details of the physics of bell, read *The Theory of Sound* by Rayleigh [9].

Coffee Break

What happens when $m = 0$, or $m = 1$?

   "$m=0$" is a motion of the bell itself around the bell axis (therefore no nodal line), and "$m=1$" is a swinging motion in which there exists one meridian pair as a node in the direction of its swinging. These motions do not create tones.

   KT has dreamed to strike a "soundless" bell. She stroke a bell with full power. Nevertheless the bell was silent.

8.2.2.2 Buddhist Temple Bells

The shape of bells in Buddhist temples is different from European bells (see Fig. 8.19). European bellfounding had made efforts to obtain good sound/harmony and finally obtained a proper shape and material. Many of us wonder, then, what is special about Buddhist temple bells.

There are three kinds of tones, when a bell is struck. They are called "atari", "osh" and "okuri" in Japanese. "Atari" corresponds to the tap (strike) tone, which disappears within 1 s. "Oshi" is a tone with high pitch which propagates further, lasting about 10 s. This is a mixture of overtones which are most probably not in harmony. "Okuri" is the deepest tone (hum), lasting from 30 s to 1 min. Moreover, beats are important. Many Japanese people feel comfortable to hearing beats with a frequency ≈ 1 s. A slight difference of the thickness and imperfect form (not exactly axially symmetric) cause vibrations of two frequencies which are close to each other, resulting in such beats.

In Japan there is a tradition to strike a temple bell 108 times on New Years Eve. 108 earthly desires are cleaned with each strike and with the last tone people go into the New Year. For this purpose, "okuri" is the most important sound which gives enough time to think about each earthly desire.

Here we see a cultural difference between Europe and Japan (or East Asia). For European bells harmony is most essential. Bells with good harmony sound good for European people. Buddhist tempel bells create inharmonic sound and beats. Japanese enjoy "controlled" inharmonicity and beats for their quiet soul.

European bells are usually struck many times, one after the other. If bells are not tuned well, they create uncomfortable dissonances. On the other hand, Buddhist temple bells are struck only once in a while. Slight dissonances and beats are even attractive.

8.3   The Human Voice: Singing

8.3.1   Tonal Range of Singers

When singing a tune, our voice serves as a musical instrument with rather special properties. There are six voice types in the operatic systems of classification (see Fig. 8.22):

- soprano: the highest female voice, being able to sing C4 to C6, and possibly higher
- mezzo-soprano: a female voice between A3 and A5
- alto: the lowest female voice, F3 to E5. Rare contraltos possess a range similar to that of the tenor.
- tenor: the highest male voice, B2 to A4, and possibly higher
- Baritone: a male voice, G2 to F4
- Bass: the lowest male voice, E2 to E4

Note that the ranges given are approximations and there are some individual deviations.

8.3.2   Speech Organs

Together with the evolution of the brain and the auditory system humans developed their speech organs (articulation organs) to a very high quality. The human voice is created by oscillation of vocal cords (also called vocal folds,

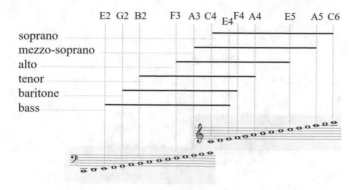

Fig. 8.22 Approximate tonal ranges of soprano, mezzo-soprano, alto, tenor, bariton and bass

located in the larynx) and modulated in the mouth, pharynx (part of the throat) and nasal cavity (see Fig. 8.23). The vocal cords vibrate due to the principle of the pressure controlled valve (see Sect. 7.5.1). The voice is used for delivering information in form of speech or other sounds like singing, crying, weeping, laughing or groaning. On the other hand, snoring, whistling or teeth grinding do not belong to the voice. Vocal cords are not involved for creating these "noises".

Sounds for speech can be divided into two groups, "voiced" sounds and "unvoiced" sounds. The voiced sounds are created by vibration of the vocal cords, while the unvoiced sounds are created by using lips, teeth or the tongue without oscillating vocal cords. All vowels (a, e, i, o, u, and more in other languages) are voiced. Some consonants are voiced and others unvoiced. For example, in English b, d, g, v, z belong to voiced sound, and p, t, k, f, s, h are unvoiced. "th" is voiced or unvoiced depending on what comes next. Whether sounds are voiced or unvoiced depends also on languages. In German "s" is often pronounced as a voiced sound (z), and additionally there are cases to be pronounced as two different unvoiced sounds (s and sh), depending on what comes next.

For singing the human body is used as a musical instrument to create melodic sound (in most of cases) with text. The tonal ranges strongly depend on the length of vocal cords. The length of the male vocal cords is between 17 and 25 mm, while that of the female's is between 12.5 and 17.5 mm. Different from the instruments, however, the acoustic characteristics of the mouth-nose-throat space can be changed by the speech organs: for example, form of the mouth, displacement of the tongue, shifting the soft palate, form of the glottis (the opening between the vocal cords), and so on, and therefore, various techniques have been developed for classic singers.

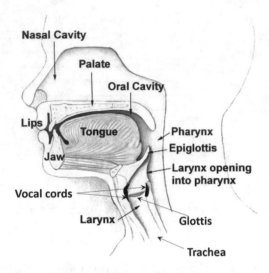

Fig. 8.23 Speech organs (public domain by Arcadian)

8.3.3 Formant

The concept of harmonic series has been intensively discussed in Chap. 4. The voiced sound also includes such a series in addition to the fundamental.

A formant is the concentration of frequency components with certain intensities in a harmonic spectrum. (This term was first used by the German physiologist Ludimar Hermann [16] in 1890). They occur, for instance, in the spectra of musical instruments as exemplified for the oboe or the horn (see spectra in Fig. 7.11). They are also characteristic for the human voice—some frequency domains are enhanced in relation to others. In the resonating body of a musical instrument or on the way between larynx and mouth opening, starting with the production of a fundamental tone, a part of the harmonic spectrum and noisy components will be damped or otherwise increased. Voices and instruments frequently show several formant regions not necessarily being directly connected.

Formants are numbered from lowest frequencies to highest: F1, F2 and so on. Formant locations of the human voice for each phonetic unit are distinct and depend on the shape of the vocal tract[3] and tongue position. In Table 8.2 the first two formants of some vowels at speech are listed [17].

For speech the first and the second formants (F1 and F2) are important to distinguish, which vowel is spoken. However, a formant map (plots of frequencies of the first formant (F1) vs. the second formant (F2)) indicates

---

[3]The human vocal tract is the cavity in human beings where the sound produced at the larynx is filtered.

Table 8.2 Average frequency of formants of some vowels

Vowel (IPA)[a]	Formant F1 (Hz)	Formant F2 (Hz)
i	240	2400
e	235	2100
a	850	1610
o	360	640
u	250	595

[a]IPA: International Phonetic Alphabet

some deviations and overlapping [18]. Many studies on how babies learn to distinguish vowels instead of the overlapping are reviewed by Kuhl [19]. The higher formants (F3, F4, ...) are not essential for understanding speech, but play an important role to characterize the anatomy of the speaker, as well as the articulation properties of the speaker, resulting in the timbre [20]. This is the reason why each human has a characteristic voice, which can be differentiated among each other.

Now we come back to the singing voice. Figure 8.24 illustrates spectral envelops for vowels u, o, a, e, and i, where a trained classic singer sings with the fundamental tone 128 Hz ($\approx$C3). The form of the spectral envelope does not vary so much, when the same person sings with different fundamentals [21]. Different from speech, however, the formants around 3000 Hz are as intensive as lower formants. Such intensive formants are absent in speech or in the spectra of untrained singers [22]. As a result songs of trained classical singers transcend the orchestra.

The frequency location of dominating formant occurs around 2300–2500 Hz for a bass, 2500–2700 Hz for a baritone and 2700–2900 Hz for a tenor, depending on the length of the vocal tract. For the female singers the frequency location is accordingly higher.

8.3.4 Singers

At the end of this chapter we would like to mention some prominent singers in classic music without further comments. .... but we cannot finish the list, sorry........

- Enrico Caruso (1873–1921)
- Maria Callas (1923–1977)
- Dietrich Fischer-Dieskau (1925–2012)
- Lady Janet Abbott Baker (1933–)

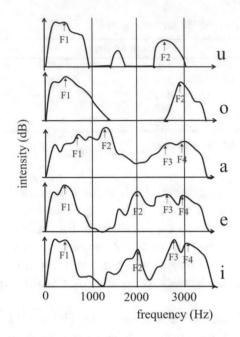

Fig. 8.24 Spectrum envelops of vowels u, o, a, e, and i (from top to bottom) at the fundamental frequency 128 Hz sung by a classic singer; data taken from [21]

- Luciano Pavarotti (1935–2007)
- José Plácido Domingo (1941–)
- Lady Kiri Jeanette Claire Te Kanawa (1944–)
- Josep Maria Carreras (1946–)
- Cecilia Bartoli (1966–)
- Jonas Kaufmann (1969–)
- Anna Yuryevna Netrebko (1971–)
- ....
- ....

References

1. W. Bray (ed.), *Memoirs, Illustrative of the Life and Writings of John Evelyn* (Henry Colburn, London, 1819)
2. H. Helmholtz, *Die Lehre von des Tonempfeindungen, als physiologische Grundlage für die Theorie der Musik* (Friedrich Vieweg und Sohn, Braunschweig, 1863)
3. L. Cremer, *Physik der Geige* (Hirtzel Verlag, Stuttgart, 1981)

4. D. Ullmann, *Chladni und die Entwicklung der Akustik von 1750–1860* (Birkhäuser, Basel, 1996), pp. 163–171

5. C.M. Hutchins, K.A. Stetson, P.A. Taylor, Clarification of "free plate tap tones" by holographic interferometry. J. Catgut Acoust. Soc. **16**, 15–23 (1971)

6. E.F.F. Cladni, *Entdeckungen über die Theorie des Klanges* (*Discovery on the Sound Theory*) (Weitmanns Erben und Reich, Leipzig, 1787)

7. H. Crew, A. de Salvio trans. *Galilei, Galileo* (Macmillan, New York, 1914) (first published in Italian 1638), pp. 101–102

8. H. Robinson, W. Adams (eds.) *The Diary of Robert Hooke, M.A., M.D., F.R.S., 1672–1680* (Taylor & Francis, London, 1935), p. 448

9. J.W.S. Rayleigh, *The Theory of Sound*, vol. 1 (Macmillan and Co., London, 1877)

10. T.D. Rossing, G. Kvistad, Acoustics of timpani: Preliminary studies. The Percussionist **13**, 90–96 (1976)

11. R.I. Mills, The timpani as a vibrating circular membrane: an analysis of several secondary effects, M.S, thesis, Northern Illinois University, 1980

12. R.S. Christian, R.E. Davis, A. Tubis, C.A. Anderson, R.I. Mills, T.D. Rossing, Effects of air loading on timpani membrane vibrations. J. Acoust. Soc. Am. **76**, 1336–1345 (1984)

13. J. Meyer, *Acoustics and the Performance of Music* (Springer Science and Business Media, Heidelberg, 2009)

14. T.D. Rossing, I. Bork, H. Zhao, D. Fystrom, Acoustics of snare drums. J. Acoust. Soc. Am. **92**, 84–94 (1992)

15. S. Shen, Acoustics of ancient Chinese bells. Sci. Am. **256**, 94 (1987)

16. L. Hermann, Beiträge zur Lehre von der Klangwahrnehmung. Pflug. Arch. Ges. Phys. **56**, 467–499 (1894)

17. J.C. Catford, *A Practical Introduction to Phonetics* (Oxford University Press, Oxford, 1988), p. 161

18. J. Hildenbrand, L. Getty, M. Clark, K. Wheeler, Acoustic characteristics of American English vowels. J. Acoust. Soc. Am. **97**, 3099–3111 (1995)

19. P.K. Kuhl, Early language acquisition: cracking the speech code. Nat. Rev. Neurosci. **5**, 831–843 (2004)

20. E. Thienhaus, Neuere Versuche zur Klangfarbe und Lautstärke von Vokalen. Zeitschr. Techn. Phys. **15**, 637 (1954)

21. E. Meyer, E.-G. Neumann, *Physikalische und Technische Akustik* (Friedrich Vieweg und Sohn, Braunschweig, 1974), p. 229

22. J. Sundberg, The acoustics of the singing voice. Sci. Am. **236**, 82 (1977)

Part IV
When Musicians Get Together, They…

For others composition is mission, work, obligation,
it signifies the entire existence;
for me composition is the calm, joy, a mood,
that diverts me from my official obligations as a professor.

—*Alexander Borodin*

9

Create Music

9.1 Introduction

When naming composers, we mention Bach, Händel, Haydn, Beethoven, Mozart, Schubert, and, and..., and we think of, at most, 10 to 20 eminent persons. Therefore, when KT read the *Musicophilia* of Oliver Sacks [1], in which his patients are often composers, she wondered whether there exist so many of them.

On the other hand, if musicians are asked to name physicists, they may list Archimedes, Galilei, Newton, Maxwell, Einstein, Marie Curie, Feynman, Hawking, and, ... They may also count 10 to 20 renowned names. When SCM introduced himself as a physicist in front of young participants of a music competition, they might have wondered whether he is really a physicist. They wonder whether there are so many physicists.....

In this world there were and are many composers and many scientists, as well. Some are famous, while others are not. Nobody is sure which names among currently active musicians or scientists will be remembered after 100 years. Our neighbor could be a composer or a physicist, though he/she is not so great as Bach or Einstein. So, why do we, as scientists, not try to learn a little bit of composition?

If we think about the ancient Greeks, music belonged to the the quadrivium together with arithmetics, geometry and astronomy [2]. Music is a function of time: frequency and sound pressure are variables of this function. To construct a piece from smaller elements, logical thinking is involved. Therefore, as a physicist or natural scientist composing musical pieces may be a field which is not so far from their daily research works.

© Springer Nature Switzerland AG 2021
K. Tsuji et al., *Physics and Music*,
https://doi.org/10.1007/978-3-030-68676-5_9

However, we have to go further. To create impressive music there is something more important than composition technique. Why are some musical pieces liked by many people and others not? It depends certainly on the character of the music. Listeners often expect melodies, which are somewhat familiar to them (maybe they have heard analogous melodies before) and which are easy to remember, maybe because of their repetitive character. Also they should be comfortable and satisfying. Think about children songs. Their range of pitch spans usually less than an octave, or even much less, so that it is easy to sing them. Many melodies of Mozart's works are amazingly simple, so that we recognize them again immediately, if we hear them for the second time.

In this chapter we explain, at first, various European music forms as a basic knowledge required for composing. Music is constructed with the smallest unit "motif", from which higher levels are created step by step (phrase → passage → piece → cycle). In the next section we will learn about two traditional methods for composing: counterpoint and harmony. Using a very simple example we will show how a canon and a fugue are made, and how the music ends safe and sound. In the last section we will discuss the choice of appropriate instruments, depending on the pieces.

9.2   Music Forms

9.2.1   Levels of Musical Units

Before learning the concepts of composition, one should know about forms and structures of music (refer to Spring and Hutcheson [3]). We will start with several units of musical notes and their levels (see Fig. 9.1). For this purpose we use the term "piece" as a subordinate term for "musical work". (Note that generally "piece" has a wider meaning referring to any musical composition.)

The smallest structural unit is called "motif", corresponding to "word" in languages. A motif consists of usually 2 bars (sometimes even of only 1 bar) and contains an essential idea of the music. Below two examples of motives are drawn. The first one is the one-bar motif which appear in many of Bach's works. The second one is the first four notes of Beethoven's Symphony No. 5 in C minor, Op. 67. This four-note motif is called "Schicksals (Fate) Motif", and the symphony is known under the name "Schicksals-Sinfonie" (Fate Symphony), though this name did not originate from Beethoven himself.

Left: Bachs's motif, right: Motif of Beethoven's Fifth Symphony

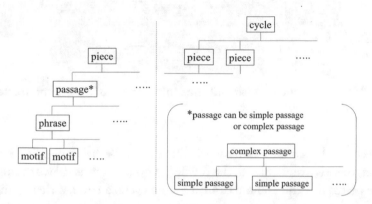

Fig. 9.1 Levels of the musical units. Left: structure of a stand-alone "piece", top right: some pieces are collected to form a "cycle". The highest level is either a cycle or a piece

The next level is the "phrase", which corresponds to a "sentence" or "verse" in language. A phrase is the fundamental unit with a definite rhythm and melody, as well as harmony, and often consists of 2 motives (four bars).

The third level, "passage", is a coherent group of several phrases, corresponding to "paragraph" in literature. There are a simple passage and a complex passage. A simple passage is a group of phrases, while a complex passage includes several simple passages (see Fig. 9.1 bottom right). Sometimes, the passage is also called a "section".

And some passages together will define a next level, a "piece", which can be compared with, for instance, an "essay", or an "article" in literature. For practical reasons we use the lower-case letters a, b, c, ... solely for the simple passages, and the capital letters A, B, C, ...for passages which are direct elements constructing a piece. It means that in some cases a = A, and in other cases A ∈ (a, a', ...). In the latter case A contains some simple passages.

We will show a simple example of a piece: the German children song "Morgen kommt der Weihnachtsmann" (Tomorrow comes Santa Claus). This piece consists of 2 (simple) passages, where the identical notes are repeated as the second passage. Each passage has 3 phrases, where the first and the third phrases are identical. Each phrase contains 2 motives (see Fig. 9.2).

Fig. 9.2 "Morgen kommt der Weihnachtsmann": example of a piece constructed from motives, phrases, passages

A piece can be a stand-alone work like a song, waltz, polonaise, or other. However, there is a higher level: the "cycle". A cycle is a musical work consisting of some pieces together. These can be a song cycle, a suite, a theme and variations, a sonata, a symphony, a concerto, an opera, an oratorio, a ballet or an incidental music. Note that in the case of sonatas, symphonies and concertos pieces making them up are called "movements".

9.2.2 More About a Piece

Above we have described that some passages together will become a piece. In this subsection we explain various ways to combine passages, A, B, ... They are the direct elements for the next level, pieces.

In Table 9.1 the structures of various pieces are summarized.

9.2.3 More About a Cycle

Individual pieces that aggregate into larger works are considered as cycles. Below we will explain typical cycles.

9.2.3.1 Suite

A suite is an ordered set of pieces. In Baroque music suites were popular and contained several dance pieces in the same key. A good example is the French Suite No. 1 in D minor BWV 812 of Johann Sebastian Bach, consisting of

Table 9.1 Various pieces and their structure

Name	Structure	Explanation
Strophic	AAA...	A song structure in which all verses or stanzas of the text are sung repeatedly, also called verse-repeating form or chorus form
Medley	ABCE..., AABBCCDD..	An indefinite sequence of self-contained sections, also called chain form
Binary	AB, AABB, AA'BB'	A form consisting of two passages A and B. Usually a minuet has the form of AABB
Ternary	ABA, ABA', ABC	A form consisting of three passages (2 times A and B), (A, A' and B) or (A, B and C)
Trio	ABA	This is also a ternary form, but each passage is itself either in binary or simple ternary form
Rondo	ABACABA, ABACADAEA	A recurring theme A alternating with different passages B, C, D,..., either symmetric or asymmetric. A may be varied to A' or A''....
Variation	AA'A''A'''...	A theme A repeated in an altered form. The changes may involve melody, rhythm, harmony, counterpoint, timbre, orchestration or any combination of these
Sonata	(introduction) exposition development recapitulation (coda)	Usually the exposition part presents a first (main) and a second theme. The second theme has often a key dominant to the key of the first theme (the tonic key), and a contrast style. This exposition is then "developed". In the recapitulation, both themes appear again in the tonic key
Sonata rondo	ABA-C-ADA	A blend of sonata form and rondo form

6 dance pieces, Allemande, Courante, Sarabande, Minuet I, Minuet II, and Gigue.

Later, in the 19th century, suites were constructed by extraction from ballet or opera music, like the Nutcracker Suite of Peter Ilyich Tchaikovsky, the Carmen Suite of Georges Bizet, the Daphnis et Chloé Suite of Maurice Ravel.

Another noteworthy work in this category is "The Planets" by Gustav Holst. (This is a large-scale orchestral work, and therefore we are not sure, whether we can call it "suite".) He expressed each planet with characteristic melody, rhythm and harmony.

9.2.3.2 Song Cycle

A lieder cycle (German: Liederkreis or Liederzyklus) is a group of individually complete songs designed to be performed in a sequence as a unit.

Franz Schubert created two song cycles: "Die schöne Müllerin" (The Fair Maid of the Mill) and "Winterreise" (Winter Journey), the poems of which were written by Wilhelm Müller (1794–1827). One more cycle "Schwanengesang" (Swan Song) was collected posthumously. Not only Schubert, but also Carl Maria von Weber, Robert Schumann, Johannes Brahms, Louis-Hector Berlioz, Modest Mussorgsky, Vaughan Williams, Arnold Schönberg, Dmitri Shostakovich, and many others wrote song cycles in various languages.

9.2.3.3 Theme and Variations

When variations are set for an independent cycle, it is usually called "Variations on the theme by X" or "X variations". (X is the name of the composer or the theme.) Examples are 12 Variations on a French lied "Ah, vous dirai-je, maman" in C major KV 265 of Wolfgang Amadeus Mozart, Variations on a theme by Joseph Haydn in B♭ major Op. 56a of Johannes Brahms.

9.2.3.4 Sonata

The word "sonata" can be traced back to the Latin word "sonare", which means "to sound". This is the antonymous word to "cantata" (in Latin "cantare" = "to sing"). So, a sonata is a work played with instruments. It consists of two to four pieces (called movements) and at least one movement has a sonata form. In the classic music sonatas are composed mainly for solo playing with keyboard instruments, duo or trio with a piano, or trio, quartet, quintet for strings.

Haydn wrote many piano sonatas with two movements. However, soon after the common structure of a sonata became to contain three movements:

- the first movement: usually allegro in sonata form
- the second movement: frequently andante, adagio or largo, in other cases, in the form of minuet or theme and variations
- the third movement (Finale): often in form of minuet or rondo in allegro or presto

Sonata No. 16 in C major, KV 545 of Mozart has the typical form mentioned above, consisting of Allegro, Andante and RONDO. There are also many variations for the structure of the three movements. For example, "Mondscheinsonate" (The Moonlight Sonata, Piano Sonata No. 14 in C♯ minor Op. 27, No. 2) of Beethoven starts with the first movement in Adagio sostenuto. The second movement is Allegretto and the third movement Presto agitato.

Different from sonatas for piano (or for piano and other instruments) the four-movement sonatas are standard for string instruments. The first and the last (4th) movements are similar to those for piano sonatas, while the second movement is frequently Andante, Adagio or Largo in sonata form (in other cases, in a form of minuet or theme and variations), and the third movement is often in a dance form like minuet, scherzo.

Haydn's famous string quartet Opus 76, No. 3 ("Emperor") contains four movements—Allegro, Poco Adagio, cantabile (this part is "theme and variations"), Menuetto, Allegro and Finale Presto. (More on this quartet see Chap. 13.)

9.2.3.5 Symphony

A symphony is a work composed for orchestra. It usually has a structure of four movements similar to sonatas for string instruments.

- the first movement: usually allegro in a sonata form
- the second movement: frequently andante, adagio or largo in sonata form, in other cases, in form of minuet or theme and variations
- the third movement: often in dance form like minuet, scherzo
- Finale: often in a sonata or sonata-rondo form with allegro-vivace, presto...

9.2.3.6 Concerto

A concerto is a musical composition generally composed of three movements, in which either one solo instrument (a piano, violin, cello or flute) or a group of soloists (in this case it is often called "concertino") is accompanied by an orchestra (Fig. 9.3). The structure is similar to that of sonatas.

Different from a symphony, there is a passage in which the orchestra stops playing and the soloist plays alone. This passage is called "cadenza". How long and what the soloist plays is originally free. But many soloists play a written passage which is often created by somebody other than the original composer. Of course, it can be improvised. There are also cadenzas which are written by the original composers. For example, Beethoven's Piano Concerto No. 5 in E♭ major, Op. 73 (the "Emperor" Concerto) contains a notated cadenza. It is specified that the soloist should play only the music written in the score and not add a cadenza on his own.

Fig. 9.3 David Oistrakh playing a violin concerto in the German National Opera, 1960 (CC BY-SA 3.0 de, Bundesarchiv, Bild 183-77066-0002 by C. Hochneder and I. Eckleben, Ir)

9.2.3.7 Opera

The opera belongs to the theater, in which actors in the stage costume perform their roles mainly by singing. An orchestra accompanies them. Actors are singers.

The opera has a longer history than the symphony, starting at the end of the 16th century: The earliest known work is Dafne by Jacopo Peri (around 1595). Afterwards operas developed in various types: an "opera seria" (a serious opera) and "opera buffa" (a comic opera) in Baroque operas, the bel canto opera movement (by Rossini, Bellini, Donizetti, Puccini, Mercadante and many others), and reforms by Christoph Willibald Gluck and Mozart to German language operas. Richard Wagner named his operas a "complete work of art". There are also French operas, English operas, Russian operas and other national operas.

We like to mention one opera, in somewhat different style: "Boris Godunov" by Modest Mussorgsky (composed between 1868 and 1873). Mussorgsky prepared not only music, but also his own libretto (text) according to Pushkin's drama. It was rejected many times for a premiere. But finally, as César Cui wrote, "The success was enormous and complete; never, within my memory, had such ovations been given to a composer at the Mariinsky" (see Fig. 9.4).

Fig. 9.4 The Death of Boris from the premiere production of the opera Boris Godunov at the Mariinsky Theatre in 1874 (public domain, from the Grove Dictionary of Music)

An opera starts with an overture, before the curtain is raised. The whole opera is usually divided into two or three acts, and there is an intermezzo. Each act includes arias, recitatives, duets and choruses. They are accompanied by an orchestra.

- Overture: In early times (the 16th–17th century) the form of the overture was either binary (AB) or ternary (ABA). After Gluck's reform in the 18th century overtures became free from these forms and foreshadowed the plot and corresponding music. Usually the overture is played by the orchestra without chorus or soloists.
- Intermezzo (Interlude): An intermezzo is an orchestral passage between acts or scenes. The intermezzo of the opera "Cavalleria rusticana" by Pietro Mascagni (written in 1889) is one of the most liked intermezzi.
- Aria: It developed from strophic to binary and then ternary form. Similar to the case of the overture, later through the 19th century, arias became free from the form, because to narrate the story is more important. Arias are often performed as stand-alone pieces.

- Recitative: This is the part where a singer imitates the rhythm of the spoken word. Usually there is no melody in the recitative passages, which are sung on a repeated note or just a few notes.
- Duet and more: Duets are performed to show a dialogue of friendship, love, confrontation, arguments, and others. There are cases in which more than two persons sing together. One famous extreme is a sextet in the opera Don Giovanni of Mozart.
- Chorus: The choruses represent groups such as soldiers, people in a village, or others which appear in the story.

An operetta is a kind of small opera (or an "opera-light"), usually shorter than operas. It originates from the opera buffa. Different from the opera, dance is also an important element of the operettas, though singers themselves do not necessarily dance. Most of the operettas are amusing and have a happy end. Famous compositions are "Orphée aux Enfers" (Orpheus in the Underworld) by Jacques Offenbach (premiere in 1858) and "Die Fledermaus" (The Flittermouse) by Johann Strauss II (written in 1874). The overtures of these operettas are often played in new year concerts.

A "musical" is a theater form in which music, songs, conversations and dances are combined. Actors should be able to sing as well as to dance. Originally the word "musical" is the short form of musical theater, and it includes musical plays, musical comedies and musical reviews. The music used is much wider than that used in an opera: it can be pop, rock or folk music. The form can be similar to an opera, but most of them deviate from it: there is more freedom. Some musicals have become popular all over the world, for instance, "Porgy and Bess" of George Gershwin, which plays an important role of transition from the opera to the musical, or "West Side Story" by Leonard Bernstein.

Coffee Break

Johann Strauss II (1825–1899) is more famous as "The Waltz King" than as the composer of "Die Fledermaus". But why is he called Johann Strauss II? (Do not mix up Johann and Richard: for Richard, see Chap. 11.) His father, Johann Strauss I (1804–1849) was also a composer of waltzes and his best known piece, the Radetzky March. Nevertheless or therefore, he forbade his son to become a musician. It was very good for us that Strauss I left his family for a mistress. After that Strauss II was able to become active in music. Otherwise we would not have had any chance to listen to his waltz "An der schönen, blauen Donau", Op. 314 (On the Beautiful Blue Danube). Johann Strauss III (1866–1939) was a nephew of Strauss II, who was rather more active as a conductor than a composer.

9.2.3.8 Oratorio

An oratorio is a large cycle composed for orchestra, chorus, and soloists. Other than the opera an oratorio is not theater but pure musical work. In an oratorio the chorus plays a central role. The plot deals with sacred topics and, therefore, it is often performed in the church. The musical form became popular in Italy in the early 17th century.

Oratorios usually contain an overture, arias, recitatives, and choruses. The instruments often include timpani and trumpets.

An early significant work is "Jephte" by Giacomo Carissimi (written around 1650). Later Georg Friedrich Händel (in England he used the name "George Frideric Handel") composed the "Messiah" in 1741. Felix Mendelssohn Bartholdy wrote "Elias" (Elijah) Op. 70, MWV A 25 in 1846.

9.2.3.9 Ballet

A ballet is the theatrical form of dance, originating in France during the 17th century. Ballet dancers perform on a stage while accompanied by music. Prior to Tchaikovsky the ballet musics were rather simple, just easy to dance, and they were even not composed by symphonic maestros. "The Swan Lake" of Tchaikovsky is the first symphonic music for ballet. He also wrote "The Sleeping Beauty" and "The Nutcracker" (Fig. 9.5). Since then ballet music is also performed without ballet dancing. Maurice Ravel composed some ballet musics, too. "Daphnis et Chloé" is one of them. Boléro was originally composed for a ballet.

9.2.3.10 Incidental Music

According to the Webster English Dictionary an incidental music is a descriptive music played or to be played during the action of play to heighten a situation or project a mood (as of a battle, a storm, a death scene…) or to relate directly to stage action (as a song or a dance). It should not be mixed up with a background music, which is specially composed or arranged to accompany the dialogue or action (but not dancing or singing) of a film, radio or TV drama.

Egmont, Op. 84 by Ludwig van Beethoven is a set of incidental music pieces for the play "Egmont" by Johann Wolfgang von Goethe. Franz Schubert wrote an incidental music "Rosamunde, Fürstin von Zypern" (Rosamunde, Princess of Cyprus) for a play by Helmina von Chézy. "Ein Sommernachtstraum" of Felix Mendelssohn Bartholdy is another good example. He composed music

Fig. 9.5 A performance of Snow Dance in Nutcracker in 1981 (CC BY-SA 3.0, poto by Rick Dikeman)

for William Shakespeare's play "A Midsummer Night's Dream" at two separate times. The first one is an overture for piano for four hands which he and his sister Fanny played together. Soon after, in 1826, he rearranged it for orchestra (Op. 2). In 1842 he wrote the incidental music Op. 61 for Shakespeare's entire play. The overture was also incorporated. Only a few years later he died.

9.3   Methods

9.3.1   Counterpoint

The word "counterpoint" is derived from Latin "punctus contra punctum", meaning "a (music) note against another note." The counterpoint was established in the 16th century. Previously, there was only a monophonic melody.

In a counterpoint more than 2 melodies are overlaid. There are neither a main melody nor subordinate melodies, but they are equivalent. An essence of the counterpoint is the beauty of overlaid sound. In other words, overlaid tones should not clash with each other. Therefore, there are many complicated rules.

Johann Sebastian Bach is a master of this method. The best example is "Die Kunst der Fuge" (The Art of Fugue) BWV 1080, which was published after his death. This work consists of 14 fugues and 4 canons in D minor. The last fugue (No.14) is an unfinished fugue: it was suddenly stopped at the measure 239. Carl Philipp Emanuel Bach (the second son of Johann Sebastian Bach) wrote on this last sheet "Über dieser Fuge, wo der Name B A C H im Contrasubject angebracht worden, ist der Verfasser gestorben." (At the point where the composer introduces the name BACH in the countersubject to this fugue, the composer died.)—shown in Fig. 9.6.

The "BACH" signature (the sequence of the notes B–A–C–H) appears in the third part of his last Fugue. A written note is shown in Fig. 9.7.

Probably because Bach's compositions were the highest art for the counterpoint, other musicians/composers at that or at later time should look for methods other than the counterpoint in order to avoid poor replications.

Fig. 9.6 The final page of Contrapunctus No. 14, 4-voice triple fugue (public domain, from Berlin State Library, Germany)

Fig. 9.7 Top: Johann Sebastian Bach (aged 61) in a portrait by Elias Gottlob Haussmann, copy or second version of his 1746 canvas. The original painting hangs in the upstairs gallery of the Altes Rathaus (Old Town Hall) in Leipzig, Germany, (public domain). Bottom: part of the written notes of the unfinished fugue, where the sequence of notes B, A, C, H appears

> **Coffee Break**
>
> We are sorry for the English notation B♭—A–C–B. Bach asks "Wat is dat?".

However, for the people who like to be composers it is even now very important to learn the counterpoint. We explain this method briefly with a very simple example.

9.3.1.1   Canon

Let us consider the first four bars of the first line in the following sheet as a unit of song. The leader sings this melody repeatedly. The second person starts singing the same melody starting from the 3rd bar, the third person from the 5th bars. They repeat the melody several times, till the leader stops singing. The second and the third persons' singing ends after 2 bars and 4 bars, respectively. This is the simplest kind of canon, and also called the perpetual/infinite canon.

Kinko's canon. The upper line is sung by a leader, the second line by the first follower, and the third line by the second follower. "l r l r l r j, j, "means left, right, left, right, left, right, jump, jump on the trampoline.

While reading the sheet, just imagine that a boy starts jumping on a trampoline: left, right, left, right, left, right, jump, jump, ..., ("Join me" to call another boy, and he joins), (2 boys jumping) left, right, left, right, left, right, jump, jump, ..., ("Join me" to call the other boy, and he joins), (3 boys jumping) left, right, left, right, left, right, jump, jump, left, right, left, right, left, right, jump, ...

If three boys jump synchronously, they can continue jumping without crash till they are exhausted.

If these boys are creative, they try something different, maybe more interesting... One of the boys may jump with handstand, the other may slow down the tempo twice.

This kind of variations results in more complicated canons. For example, the second person sings with higher pitch, may be by the interval third. In an inversion canon, the second person sings in motions contrary to the leader: If the leader goes down by a particular interval, the second person goes up by the same interval. In the case of a crab canon the second person sings backward. Also there exist canons where the rhythm of the follower is an integral multiple of others.

9.3.1.2 Fugue

Originally any kind of imitative musical counterpoints were called fugues. However, since the 16th century, the "canon" has been split off from the "fugue". Different from the canon, there is in a fugue no strict imitation, but some variations from the main melody are allowed.

Below we show a simple 4-voice fugue, the main melody of which is analogous to the canon above. Different from three boys in the canon, we have here actions of a family with 4 members. A girl starts jumping on a trampoline alone, left, right, left, right, left, right, jump, jump, ... , at the middle of the 4th bar, the mother, who is heavier than the daughter, joins and they jump together in different rhythms. At the 9th bar the father takes over the mother, and the elderly brother joins his sister. At the middle of the 11th bar, mother joins again, and all four are jumping on the trampoline.

Let us hope that the 4 members can enjoy trampoline jumping without any hard crash.

Fig. 9.8 Beginning of Kinko's 4-voice fugue

9.3.2 Harmony

During the Renaissance period people became increasingly aware of harmony. Even in counterpoint music the character of overlaid tones should be important. However, in counterpoint there are predominantly melodies. Overlay of sound/harmony comes after melodies. On the other hand, in the case of harmony methods there are harmonies at first, and then these harmonies are connected one after the other for constructing melodies.

Jean-Philippe Rameau (1683–1764) was a French composer and music theorist. He is called "Father of harmony" because of his great work *Traité de l'harmonie* of 1722 [4] (Fig. 9.9). He summarized various intervals with drawings of coordinates like in mathematics, and also proposed the idea to use three main functions of harmony, the tonic, dominant and subdominant, though these names do not stem from him. He put forward his theory of chord progression and cadence. His Treatise became the authority on music theory, forming the foundation for instruction in European music. (As we mentioned in Sect. 3.1.1, he created among other intervals a sixth chord (sixte ajoutée)).

Johann Sebastian Bach practiced along the Treatise of Harmony for composing "The Well-Tempered Clavier". Later, Rameau's theory parted into two ways, the German theory by Hugo Riemann [5] and the Viennese theory by Simon Sechter, Arnold Schönberg, Heinrich Schenker and others [6].

Fig. 9.9 Left: Jean-Philippe Rameau by Jacques Aved, 1728 (public domain, Collection of Musée des Beaux-Arts de Dijon), right: title page of the Treatise on Harmony (public domain)

The concepts of the German theory based on Riemann's "dualist" system are:

1. There is only one chord (a triad formed on a tonic note of a scale) to be used; this is consonant and generates a feeling of repose and balance.
2. This chord has two different forms, major and minor, depending on whether the chord is composed of a minor third over a major third, or a major third over a minor one.
3. This chord is able to take on three different tonal functions, tonic, dominant, or subdominant.

The Viennese theory is characterized by the use of Roman numerals to denote the chords of the tonal scale, as shown below (the seven tonal steps in C major with their respective triads and Roman numeral notation):

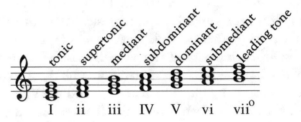

The seven scale degrees in C major denoted by Roman numbers (CC BY-SA 3.0 by Fauban)

One considers each step to have its own function and refers to the tonal center through the circle of fifth (see Chap. 4).

One nice example of harmony is shown in Fig. 9.10: the first and the last phrases of a church song "Es ist ein Ros entsprungen" (Lo, How a Rose E'er Blooming) arranged by Michael Prætorius. This is a song for 4 voices, and the harmony of the 4 voices sounds very comfortable and healthy. The main melody of the first phrase and the last phrase are identical. However, the harmony towards the end are different from the first phrase, so that the piece can reach the end safe and sound. We would like to remind you that this had been composed almost 100 years before Rameau's book was published.

Joseph Haydn, Wolfgang Amadeus Mozart and Ludwig van Beethoven composed their music faithfully according to Rameau's theory. Franz Schubert, Frédéric Chopin, Franz Liszt and Robert Schumann did not stick to the original theory any more, but deviated from it: doing free modulation, using diminished seventh or tension chords and others. After the middle of

Fig. 9.10 The first (top) and last (bottom) phrases of a church song "Es ist ein Ros entsprungen" for 4 voices arranged by Micael Prætorius. Corresponding harmonies in F major are denoted by Roman numbers

the 19th century, music of Richard Wagner, Anton Bruckner, Gustav Mahler and Richard Strauss moved further away from the traditional ways. Wagner's Tristan chord—a tetrad of F, B, D♯ and G♯—is a good example. From the end of the 19th century to the beginning of the 20th century impressionists came into fashion. One of them, Claude Debussy, reintroduced church modes and produced sensitive and colorful music.

We come back to our examples of Kinko's canon and fugue. Usually composers add a "coda" to the end of a canon or fugue to finish the piece. Here we try to use a sequence of harmonies for closing. There are three cadences possible to finish a piece.

- tonic-dominant-tonic: strong cadence
- tonic-subdominant-tonic: weak cadence
- tonic-subdominant-tonic-dominant-tonic: very strong cadence

We use the very strong cadence "tonic—subdominant—tonic—dominant—tonic". The boys can stop jumping with the following melody.

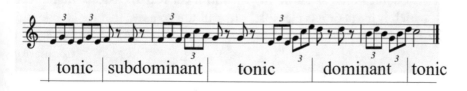

The fugue can now end as shown below:

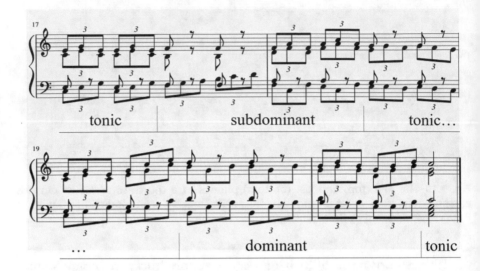

As a beginner, finally, KT thought that she composed successfully one canon and one fugue. But her work has not come to the end yet. When she tried to play the fugue on a piano, she found that some parts are rather difficult to play. We will discuss this point in the next section.

9.4   Combination of Musical Instruments

As just described, KT wrote down a melody which came to her head one morning, and constructed a canon and a fugue. Combination of tones are more or less consonant and comfortable, as far as she imagined from the sheet. She intended to write a piece for piano. However, she is still a layman, and she did not think how it can be played. The first time when she tried to play (the first version—not shown), she noticed that there are intervals which are not possible to be played at the same time. Either they are too large for her hands (intervals more than a 10th) or both hands are overlapping. So, she had to rewrite some notes, as shown in "Beginning of Kinko's 4-voice fugue" (Fig. 9.8), which is more or less playable, but still not so easy.

In order to compose music it is essential to consider with which instruments music should be played. (When professional composers read this sentence, we are sure that they say "It is too clear to be even mentioned.") Each musical instrument has specific ranges of pitch. Some instruments can be played in a polyphonic way, but others not. They have a characteristic timbre. Moreover, one has to know how the instruments are played, and what is technically possible.

Kinko's canon has a vocal range from B3 to C5. Therefore, this can be sung by children, most of women and men who are a tenor. Not only with a piano or organ, Kinko's fugue may be sung with soprano, alto, tenor and bass, though some rearrangements for certain parts are necessary. It can be also played as a string quartet (2 violins, 1 viola and 1 cello).

To make an orchestra version we add the melody part which is one octave higher than that of the first violin. This can be played with a flute. Also the basic tones which are one octave lower than the viola and cello parts. This could be played with a double bass as well as a tuba.

We can add a clarinet and oboe which would play the melody the same way as the first violin. Even if they play unisono, you hear a more compound sound because of their characteristic timbre. Due to their harmonic pattern we can add one or two horns for the melody of the second violin and a trombone and a bassoon for the viola melody. If you like to emphasize some melody, let trumpets play such parts. Percussion instruments can put accents, as well as ornaments. They have many different timbres and create interesting harmonics.

Note: Phase portraits, which are introduced in Chap. 13, can help to make an orchestral version. One can see how each instrument fills the acoustic space, as demonstrated with an example of Rossini's Guillaume Tell Overture (Fig. 13.3).

We have just tried to show, as an example, one way for composing a musical piece as laymen. There might be a lot of other ways and better ways for doing this, though. In between, KT thought that she could compose more interesting pieces if she were not restricted by the counterpoint. Of course, we are free to choose other methods and techniques to compose. Certainly KT will try to write other pieces in the future, and wishes many physicists and mathematicians to also try and compose some pieces. Physics/mathematics and composition have many aspects in common: For both, time functions and frequency functions are essential.

References

1. O. Sacks, *Musicophilia: Tales of Music and the Brain* (Picador, London, 2012)
2. E.B. Castle, *Ancient Education and Today* (Penguin Books, Baltimore, 1969)
3. G. Spring, J. Hutcheson, *Musical Form and Analysis: Time, Pattern, Proportion* (Waveland Press, Long Grove, 2013)
4. J.-P. Rameau, *Traité de l'Harmonie* Réduite à ses Principes Naturels (Dr. Jean-Baptiste Christophe Ballard, Paris, 1722)
5. H. Riemann, *Handbuch der Harmonielehre*, 6th edn. (Breitkopf und Härtel, Leipzig, 1917)
6. R.E. Wason, *Viennese Harmonic Theory from Albrechtsberger to Schenker and Schoenberg* (Ann Arbor, London, 1985)

10

Play Together and Form an Orchestra

10.1 Introduction

An essential difference between European and non-European music is that European music uses chords and harmony a lot, while in the non-European music a solo melody is accompanied by percussions or vocal interlude: Such accompanying sounds/noises make solo music more festive, more alive or substantial, but do not contribute to form harmony.

In order to create chords and harmony European music is often played together. Of course, there are numerous solo pieces in European music, too. Many piano pieces are composed for piano solo, as well as for organ, probably because with these instruments chords can be played alone. Most of other solo pieces have an accompaniment in consideration of harmony. In this sense the solo pieces in European music are also "European".

The smallest form of an ensemble is the "duo", which is played by two musicians. Then, there are: trio, quartet, quintet, ..., chamber orchestra, and finally orchestra.

Correspondingly there are groups of vocal music: solo, duet, terzetto, quartet, ..., chamber chorus, chorus.

The orchestra is the biggest group among those who "play together". We will see how the modern orchestra has been established. There are various types of orchestras, from small size to large size, with different combinations of instruments, various seating arrangements. The roles of conductor and concert master are essential to play together. We will see what they are doing before, during, and after a performance. Finally we will learn some acoustics of orchestral instruments, relating to the seating arrangements.

© Springer Nature Switzerland AG 2021
K. Tsuji et al., *Physics and Music*,
https://doi.org/10.1007/978-3-030-68676-5_10

10.2 Solo and Duo

Solo is an Italian word meaning "alone". It applies for a piece played completely alone, or with accompaniment. A soloist is a person who plays (or sings) solo pieces. Accompanying instruments are a piano or organ, or sometimes an orchestra or chorus and in Baroque music continuo instruments.

10.2.1 Piano

Some of the large musical instruments are frequently used for solo performance: those played with a keyboard. Due to this feature chords and harmonies are created as a stand-alone instrument. There are various keyboard instruments based on different principles. An organ as well as an accordion are wind instruments, a harpsichord is a plucking string instrument, and a piano is a struck string instrument. Further, a celesta is a struck idiophone, a carillon consists of bells operated by a keyboard, and there are electric keyboards like synthesizers and digital pianos, as well.

The first string instrument struck with hammers was the hammered dulcimer (see Fig. 10.1), played since the Middle Ages in Iraq, India, Iran, and other Asian countries, as well as in central Europe. In the 17th century the clavichord was invented, the strings of which are struck vertically. Note that almost at the same time the harpsichord was developed. The harpsichord does not belong to the struck instruments but to the plucking instruments. Nevertheless, the technique to make the harpsichord contributed substantially to the later development of the piano, particularly for the most effective ways to construct the case, soundboard, bridge, and mechanical action for a keyboard [1].

Fig. 10.1 Hammered dulcimer (public domain by Dvortygirl)

Around 1720 Bartolomeo Cristofori di Francesco (1655–1731) of Padua, Italy, invented the so-called Cristofori's piano, which he named "un cimbalo di cipresso di piano e forte" (a keyboard of cypress in soft and loud). This name was abbreviated later to "pianoforte" and then simply to "piano". The Cristofori piano of 1720, which is now exhibited in the Metropolitan Museum in New York, had 54 keys of a range from C2 to F6. Johann Sebastian Bach composed The Well-Tempered Clavier, BWV 846–893 during this period. The range of these pieces is from C2 to C6. Figure 10.2 illustrates the numbers and ranges of keys.

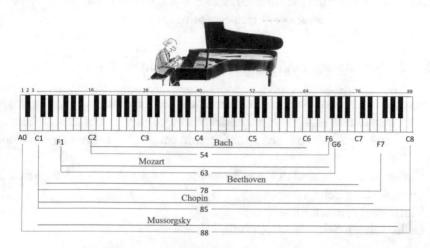

Fig. 10.2 The numbers of piano keys and the range used by composers at the corresponding time (picture on top by M. Covarrubias)

After the invention of Cristofoli's piano, music pieces explicitly for a piano and not any more for a harpsichord were composed by Muzio Clementi, Haydn and Mozart. In the Viennese Classical Era (around 1780), a lot of sonatas, as well as variations were composed. The number of keys increased to 63 (F1–G6). The minimum and maximum pitches used in the Piano Sonata F major KV 533/494 of Wolfgang Amadeus Mozart composed in 1788 are F1 and F6. In the second half of the 19th century pianos had a maximum of 78 keys (C1–F7). Ludwig van Beethoven composed his Sonata No. 30 E major Op. 109 in 1820, in which the pitch ranges from D1 to C7. Increasing the number of keys was due to the demands of composers and developments of techniques. Franz Schubert composed during this period a lot of sonatas as well as many other pieces like Fantasies, Impromptus, Moments musicaux.

In the times of Frédéric Chopin and Franz Liszt the number of keys increased to 85 (C1 to C8). Chopin's Etude Op. 10 No. 1 in C major covers notes from

C1 to E7. Many of the Romantic composers used such pianos. Towards the end of the 19th century the number reached up to 88 (A0–C8). The Great Gate of Kiev in the Pictures at an Exhibition of Modest Mussorgsky, composed in 1874, covers the range from C1 to A7.

The pitches of the piano with 88 keys range from 27.5 to 4186 Hz (when A4 is tuned to 440 Hz). Although human ears can hear tones from 20 to 20,000 Hz, it makes not so much sense to increase the number of keys further: The tones lower than C1 sound somewhat roaring and those higher than 4000 Hz are harsh noise without detailed sense of a pitch interval. Some of the piano makers construct pianos with 102 keys (C0 to B8, corresponding 16 Hz to 7902 Hz). But these are very rare exceptions.

10.2.2 String Solo Without Accompaniment

Different from the keyboard instruments, to create chords and harmony with a string instrument requires highly advanced techniques. Therefore, there are not so many occasions to listen to string solo music. Some famous works are:

- Violin: Johann Sebastian Bach 3 sonatas, 3 partitas, Niccoló Paganini 24 caprices, Béla Bartók sonata
- Viola: Paul Hindemith solo sonata
- Cello: Johann Sebastian Bach cello suites

As an exception there are solo pieces for the flute: Carl Philipp Emanuel Bach sonata, Debussy Syrinx and Prokofiev sonata.

10.2.3 Solo with Accompaniment

Many of the instruments are usually played as solo with piano accompaniment (sometimes also orchestra accompaniment). For example, Fritz Kreisler composed very sweet pieces for violin accompanied by piano: Liebesfreud (Love's Joy), Liebesleid (Love's Sorrow) and others.

There are many sonatas for a string instrument, (for example the "Sonata in A minor for Arpeggione and Piano, D 821" of Franz Schubert), sonatas with wind instruments, and with brass instruments. Remarkable is the input of Paul Hindemith, who wrote sonatas for various instruments, for example, a flute, a clarinet, a saxophone, a trumpet, a trombone, a tuba, a double bass, a harp and some more.

Furthermore, solo vocal music should not be forgotten. Franz Schubert composed many lieder with very sophisticated piano accompaniment. Sometimes, a chorus accompanies a solo singer.

10.2.4   Duo

A duo is a musical composition for two performers in which the performers have equal importance for the piece. There are works for two pianos, four hands (two persons play on one piano together), two violins or two string instruments, two wind instruments, two brass instruments, other combinations of instruments, and two singers.

There is no clear border between "solo with accompaniment" and "duo", though. A good example is found with Schubert. He composed 3 "sonatines for piano and violin". He did not name them "violin sonatines" (in which a violin is accompanied by a piano). He composed "duos for piano and violin" (Sonata in A major op. posth. 162, and Fantasy in C major op. posth. 159). It is obvious that for Schubert the roles of piano and violin were equivalent. His lieder could be considered also "duo", because a piano part plays an important role. Actually, in the International Schubert Competition Dortmund, they call a presentation for lieder "LiedDuo" competition.

Mozart's as well as Beethoven's works are also called "sonatas for piano and violin", and not "violin sonatas".

Further, there are duos with two instruments of the same kind:

- violin duo: two violins play together

    - Joseph Haydn: 6 Duos, Op. 6
    - Béla Bartók: 44 Duos, Sz. 98, BB 104
    - Jean-Marie Leclair: 6 Sonatas, Op. 3 and 6 Sonatas, Op. 12
    - Sergei Prokofiev: Sonata Op. 56

- piano duo (for 2 pianos)

    - Wofgang Amadeus Mozart: Sonata in D major, KV 448
    - Frédéric Chopin: Rondo in C major, op. posth. 73
    - Franz Liszt: Totentanz, Paraphrase on Dies Irae, 2-piano, S 652
    - Camille Saint-Saëns, Variations on a Theme of Beethoven, Op. 35

- piano four-hands

    - Franz Schubert, Fantasie in F minor, D 940
    - Felix Mendelssohn Bartholdy, Ein Sommernachtstraum, Op. 61

Coffee Break

The "piano for four hands" is more popular than the duo for 2 pianos. At first, one does not need to have two big instruments. Four-hand pieces were often played as a house music. One more nice advantage is that two players sit very close together and that sometimes the left hand of the primo (the higher part) touches the right hand of the secondo (the lower part), by chance or deliberately. Some pieces were composed in such a way that it is not possible to play them without touching the hands. Mozart and Schubert composed such pieces.

10.3 Chamber Music

The term "chamber music" [2] was originally used for music different from the "church music": The church music is played in a church, while chamber music is played in a palace chamber, "the chamber". Figure 10.3 shows a "chamber" music performed by Frederick the Great, Franz Benda, Carl Philipp Emanuel Bach and others in the summer palace Sanssouci.

Fig. 10.3 Frederick the Great plays flute in his summer palace Sanssouci, with Franz Benda playing violin, Carl Philipp Emanuel Bach accompanying on the keyboard, and unidentified string players; painting by Adolph Menzel (1850–1852), collection in Alte Nationalgalerie, Berlin (public domain)

During the Baroque period the name "chamber music" was defined as an instrumental music played with a small group. In other words, it is a form of classical music that is composed for a small group of instruments, from two to nine. The name "chamber music" is also used for a small instrumental ensemble in contrast to the orchestra, although the border between the chamber music and orchestra music is not so clear: for example, is a chamber orchestra an orchestra or a chamber ensemble?

10.3.1   Baroque Trio Sonata

The trio sonata was one of the most important categories of a chamber music during the Baroque period around 1600–1750. As we can imagine from a large number of printed notes, the trio sonatas were very much liked, not only by aristocratic but also civil as well as virtuoso music lovers. The trios are characterized by two equally ranked melody parts over the basso continuo. Operas of Claudio Monteverdi are typical examples.

Basso continuo means an uninterrupted bass, which in baroque music plays the role of a harmonic skeleton. It consists of the lowest instrumental tones with chords which are congenial to the melody and musical flows. These tones are not explicitly written as music notes on the sheets, but are specified with numbers or other symbols, so that exact performance is left to the decision of players. Therefore, there is some freedom for improvising. Modern editions, however, often include on the music sheet some possible implementations for chords.

In trio sonatas the melodic lines were more important than the tone color. Many sonatas consisted of two-part fugues, which in later works became the three-part fugues with basso continuo. Melodic parts were initially played only with violin, viola or cornet, but in the 18th century more often with wood wind instruments. Usually, the bass part was played with a cello, bass gamba, bassoon or violone.[1] The harmony guideline suggested with numbers or symbols was interpreted with a cembalo, organ or harp. Although the name is "trio" sonata, it is performed often with more than three parts. The trio sonata is also a parent of the "concerto grosso",[2] which flourished at the end of the 17th century.

---

[1] A violone belongs to either the viol or violin family. It may have six, five, four, or even only three strings.

[2] Concerto grosso is a form of baroque music in which the musical material is passed between a small group of soloists (the concertino) and a full orchestra (the ripieno or concerto grosso).

Some composers of Baroque style are Giovanni Gabrieli, Arcangelo Corelli, Henry Purcell, Jean-Baptiste Lully, Tomaso Albinoni, Antonio Vivaldi, Georg Philipp Telemann, Georg Friedrich Händel, Johann Sebastian Bach and sons.

10.3.2 String Instruments and Piano

In any ensemble all instruments should be tuned beforehand. Usually the first violin gives the concert pitch A4, and other members tune their instruments accordingly. Exceptions are ensembles with a piano. Here the piano gives the tone for tuning.

A piano trio is an ensemble consisting of a piano and two further instruments. The most common combination is a piano, a violin and a cello. The piano trio exists also in jazz, played usually with a piano, a double bass and percussion instruments (sometimes, a piano with a guitar and a double bass).

Many classic and romantic composers composed piano trios. Joseph Haydn's Piano trio No. 39 in G major (with the "Gypsy Rondo") is one of the most enjoyable pieces. Ludwig van Beethoven wrote 11 piano trios. Among them the Ghost Trio and the Archduke Trio are the most famous ones. Felix Mendelssohn Bartholdy created two beautiful piano trios. And of course, there are enchanting pieces of Wolfgang Amadeus Mozart, Franz Schubert, Robert Schumann, Johannes Brahms, Antonín Dvořák, and others.

For a piano quartet one has an ensemble consisting of a piano and three further instruments. Usually it consists of a piano, a violin, a viola and a cello. Worth mentioning piano quartets are No. 1 in G minor and No. 2 in E♭ major of Mozart and three quartets No. 1 in G minor, No. 2 in A major and No. 3 in C minor of Brahms.

Members of a piano quintet are a piano and four other instruments, usually four string instruments. Note that pieces with a piano and four wind instruments are also called piano quintet.

The four string instruments are usually 2 violins, 1 viola and 1 cello. One famous exception is the Forelle (Trout) quintet of Schubert (Fig. 10.4), which is played by 1 piano, 1 violin, 1 viola, 1 cello and 1 double bass.

Further on, there are piano sextets for which various options of instruments are chosen. For example, the Russian composer Michail Glinka composed the sextet in E♭ major with 2 violins, 1 viola, 1 cello, 1 double bass and 1 piano. Mendelssohn composed the Piano sextet in D major with 1 violin, 2 violas, 1 cello 1 double bass and 1 piano.

Fig. 10.4 Music notes of the Trout Quintet and its composer Franz Schubert drawn by Josef Kupelwieser in 1821 (public domain)

10.3.3   String Instruments Only

A string trio is an ensemble of three string instruments, typically 1 violin, 1 viola and 1 cello. Joseph Haydn, Wolfgang Amadeus Mozart, Ludwig van Beethoven and Franz Schubert are the most famous composers for this genre. Examples are: Divertimento in E♭ major of Mozart, Serenade in D major of Beethoven, String trio in B major of Schubert.

> **Coffee Break**
>
> There is the combination of 2 violins and 1 viola, too. Antonín Dvořák composed Drobnosti (Miniatures) Op. 75 a for himself and two of his students (2nd violin and viola). Drobnosti consists of four small lovely pieces, Cavatina, Capriccio, Romanza and Elegia. However, Dvořák found that for his students it was too difficult to play. Therefore, he rewrote this for violin and piano, "Romantic Pieces Op. 75" (1. Allegro moderato, 2. Allegro maestoso, 3. Allegro appassionato and 4. Larghetto). He should then have a very good student for playing piano, because the piano part is not so simple to play either. The piano part was composed from the 2nd violin to mainly the right hand of the piano and from the viola to the left hand. Maybe therefore, sometimes the right hand and left hand cover the same range of the piano keys.

A string quartet is the heart of chamber music. It consists of 2 violins, 1 viola and 1 cello. It was developed from the Baroque trio sonatas during the 18th century. The main difference between the Baroque trio sonatas and modern string quartets is the role of the cello. The cello plays not only basso continuo but also melody. Joseph Haydn founded the modern form of the string quartet around the 1750s. An early form had five movements: allegro-minuet–adagio-minuet-allegro. Later he changed the form to the four-movement sonata: moderato-adagio-minuet-presto.

Usually members sit in a half circle as shown in Fig. 10.5, so that each member can listen to other members in an optimum fashion (refer Sect. 10.5.2). A good example is the performance of the Tallin String Quartet depicted in Fig. 10.6. Sometimes the place of the viola and cello are exchanged.

Wolfgang Amadeus Mozart and Ludwig van Beethoven followed Haydn's form and established the most challenging form of the string quartet (typical succession: allegro-adagio-minuet-vivace). Franz Schubert composed among others the famous string quartet "Der Tod und das Mädchen" (Death and the

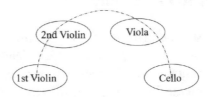

Fig. 10.5 Seating order of a string quartet

Fig. 10.6 Tallinn String Quartet performing in Tel Aviv. From left to right–1st violin, 2nd violin, viola, cello (CC BY 2.0, Estonian Foreign Ministry)

Maiden). Furthermore, Felix Mendelssohn Bartholdy, Edvard Grieg, Robert Schumann, Johannes Brahms, Bedřich Smetana, Giuseppe Verdi and Alexander Borodin are well known composers of string quartets in the 19th century. It is worth noting that Smetana included his tinnitus tone in his work, the string quartet "Aus meinem Leben" (From My Life)—more to read in Chap. 11.

In the 20th century, after the late romantic period (Maurice Ravel, Claude Debussy) some new waves came from Leoš Janáček, Arnold Schönberg, Béla Bartók, Igor Stravinsky and Alban Berg. From the point of harmony quartet became atonal, from the point of parts assignment a soprano was included (Schönbers's String Quartet No. 2), and from the point of the structure movements were not tightly connected. One example is a one-movement quartet (No. 3) of Bartók. He introduced an element of folk music, using a new playing technique, the so called "Bartók pizzicato". Here, the string is plucked vertically by snapping and rebounds off the fingerboard of the instrument, so that the sound of pizzicato is stronger and even noisy.

However, at the same time Paul Hindemith, Dmitri Shostakovich and others tried to come back to the flourishing time of Beethoven, too.

At the end of this part we would like to introduce one book with full humor about the string quartet. The book is written in German, though: *Das stillvergnügte Streichquartett* (The Contented String Quartet) by Ernst Heimeran and Bruno Aulich [3].

Furthermore, there are string quintets, sextets, septets, octets and double quartets.

Noteworthy is the Octet in E♭ major Op. 20 of Felix Mendelssohn Bartholdy, with 4 violins, 2 violas and 2 cellos (see Fig. 10.7). This work was composed in 1825 as a birthday present for his friend and his violin teacher. It is amazing that a 16 year-old boy could write such a brilliant piece, just as a Sunday music for his friends and family. A long time ago KT listened to this octet for the first time in SCM's apartment. Since then KT is thankful to SCM for introducing this wonderful music. The impact of its brilliance on KT's musical mind is the strongest so far.

There is also another form of the octet, which is a double quartet. In this ensemble two equivalent quartets play together, often alternating. Louis Spohr composed the music for such double quartets.

Fig. 10.7 Eight musicians (3 professional musicians and 5 scientists) play the Octet of Felix Mendelssohn Bartholdy. This is a symbol of "Physics and Music"! Top: from left 2nd violin, 3rd violin, 4th violin, 1st viola, 1st violin (showing the back), 2nd viola, 2nd cello and 1st cello, bottom: SCM is introducing the octet members (photo top by R. Dransfeld, bottom by T. Hashimoto)

One very special string octet is the cello octet, consisting of eight cellos. "Bachianas Brasileiras" No. 1 and No. 5 of Heitor Villa-Lobos are very popular. Beside the octet there are other variations for cellos: for example, Yale Cellos: an ensemble at the Yale School of Music consisting of about 15 cellists.

10.3.4 Together with Wind Instruments

Different from the chamber music solely with string instruments, any ensemble together with wind instruments is more complicated for developing a satisfactory systematics.

There are trio, quartet, quintet, sextet, septet, octet and nonet with various combination of wind instruments, a piano, and string instruments. Some examples are shown below:

- trio

  - 1 flute, 1 clarinet and 1 bassoon: Kontratanz und Menuett of Wolfgang Amadeus Mozart
  - 2 oboes and 1 English horn: Trio in C major Op. 87 of Ludwig van Beethoven
  - 1 violin, 1 clarinet and 1 piano: Kontraste of Béla Bartók

- quartet

  - 1 piano, 1 flute, 1 viola and 1 cello: three quartets of Carl Philipp Emanuel Bach
  - 2 clarinets 1 basset horn and 1 B-clarinet of Jean Francaix

- quintet

  - 1 flute, 1 oboe, 1 clarinet, 1 bassoon and 1 horn: Joseph Haydn, Gustav Holst, Arnold Schönberg, Paul Hindemith, György Ligeti
  - a brass wind quintet consisting of 2 trumpets, 1 horn, 1 trombone and 1 tuba. Many pieces were written as arrangements from Johann Sebastian Bach's works.
  - a clarinet quintet was in favor for many composers. It consists of a clarinet and a string quartet. Famous composers are Wolfgang Amadeus Mozart, Carl Maria von Weber, Ferruccio Busoni, Johannes Brahms, Max Reger and Paul Hindemith

- more

  - Wind sextet in E♭ major Op. 71 of Ludwig van Beethoven with 2 clarinets, 2 French horns and 2 bassoons
  - Septet for trumpet, piano, 2 violins, viola, cello and double bass of Camille Saint-Saëns
  - Introduction and Allegro for harp, flute, clarinet and string quartet of Maurice Ravel
  - Octet for flute, clarinet, 2 bassoons, 2 trumpets and 2 trombones of Igor Stravinsky

10.4 Orchestra

10.4.1 History

An orchestra is a large instrumental ensemble, which was developed in Europe.[3] Various kinds and numbers of instruments are involved, depending on the type of the orchestra.

The name "Orchestra" traces back to the old Greek $o\rho\chi\eta\sigma\tau\rho\alpha$. It was originally a place for dancing and singing (chorus) located around the altar of the Greek god Dionysos. In old Greek theaters the orchestra was an important element of the stage in an amphitheater, used for chorus and actors [4].

With the decline of Greek culture the "orchestra" was almost forgotten. After a long time in the 17th century in France, the front side of a stage, where musicians play, was called "orchestra" again. Soon after musicians or a group of musicians were called "orchestra". Thus, only about 400 years ago a large group of musicians organized with various musical instruments turned out to be called "orchestra".

The oldest opera "Dafne" composed by Jacopo Peri (1561–1633) was played in 1598 in Florence. Since then the orchestra has played an important role as accompaniment in an opera. Claudio Monteverdi (1567–1643) composed the opera L'Orfeo in 1607. For this opera he indicated, for the first time, which instruments should play which parts. This is an epoch-making work for the orchestration in music history.

Joseph Haydn (1732–1809) is called the "father of the symphony" and also the "father of the string quartet". In fact Papa Haydn composed more than 100 symphonies and more than 80 string quartets. Till about 1780 there were only four groups for string instruments (the 1st and the 2nd violin, the viola and the cello) plus the cembalo in his symphonies. From the 82th to 87th symphonies the double bass was included and string groups counted five, as is now.

The Court Orchestra in Mannheim is known as the basis for the modern orchestra. It consisted of strings, two flutes, two oboes, two bassoons, two horns, two trumpets and kettledrums. In this orchestra the strings began to coordinate bowing. During his visit to Mannheim Wolfgang Amadeus Mozart (1756–1791) was inspired very much by this orchestra.

Innovation of the musical instruments contributed to further development of orchestras in size and variety. During the time of the Cremona school (Stradivari and others, in the 17th–18th centuries) string instruments alone were able

---

[3]There are various ensembles of non-European music, too. However, here we refer to "orchestra" only in European music.

to configure the string orchestra. German horn players Heinrich Stölzel and Friedrich Blühmel invented and developed pistons and valves for horns and other brass instruments. Theobald Böhm refined the flute and developed a new fingering technique. Adolphe Sax invented the saxophone. This all occurred in the 19th century. Hector Berlioz (1803–1869) realized such benefits to compose his Symphonie fantastique in 1830.

Not to be forgotten is Ludwig van Beethoven (1770–1827). He expanded the orchestra by using clarinets, trombones, a number of percussion instruments, a piccolo and double basses. Finally he composed the famous Symphony No. 9 in D minor Op. 125 in 1824, which was the first symphony using voices.

Richard Wagner (1813–1883) built his own opera house, The Bayreuth Festspielhaus (Bayreuth Festival Theatre), opened in 1876. He could, probably for that reason, make revolutionary works for orchestration and conduction. He introduced a new idea to use a certain harmony for a leitmotif. The "Tristan chord" is an example. (This chord is used for Tristan.) For his musical dramas he expanded the orchestra. For instance, he used six harps for the opera "Das Rheingold". He also established a new style of conducting, which differs from mechanical movement for obtaining orchestral unison. In his essay "About conducting" [5] he wrote that conducting is a means by which a musical work could be re-interpreted. His approach was very flexible: tempo, dynamics and bowing of string instruments. Sometimes he even rewrote scores. Many conductors (Wilhelm Furtwängler, for example) were inspired through Wagner's ideas on conduction.

Richard Strauss (1864–1949) is considered as a leading composer of the late Romantic and early modern eras, a successor of Richard Wagner and Franz Liszt. His works for various solo instruments with orchestra are worth noting here: examples are the tone poem Don Quixote for cello, viola, orchestra, or the Duett-Concertino for clarinet, bassoon with string orchestra. The tone poem "Also sprach Zarathustra (Thus Spoke Zarathustra)" and "Eine Alpensinfonie (An Alpine Symphony)" should be mentioned.

Since the Court Orchestra in Mannheim has been established, the size of the orchestra was continuously expanding. The ultimately large orchestra was organized for the Symphony No. 8 in D major (finishes as D♭ major, though) of Gustav Mahler (1860–1911). This is also called "Sinfonie der Tausend (Symphony of a Thousand)". The orchestra consists of about 120 musicians, two choirs of a minimum of 32 singers each, 8 vocal soloists and a boy's choir (see Fig. 10.8).

Fig. 10.8 American premiere of Mahler's 8th Symphony in 1916 by Stokowski and the Philadelphia Orchestra (public domain)

Even larger is the Gurre-leader[4] of Arnold Schönberg (1874–1951). This is a large cantata which requires 20 of 1st and 20 of 2nd violins, 16 violas, 16 cellos, 12 double basses, 4 piccolos, 4 flutes, 3 oboes and 2 English horns, 7 clarinets, 3 bassoons and 2 double bassoons, 10 horns, 6 trumpets and a bass trumpet, alto trombone, 3 trombones, bass trombone, double bass trombone and tuba, 4 harps, a keyboard and many percussion instruments (for 5 persons)-totally 144 musicians. It requires also a narrator, five vocal soloists, a huge choir and a separate male choir.

However, such large orchestras are exceptions and not always present. They are organized occasionally by requirement. Usually, about 100 musicians belong to a modern standard orchestra (see below, Sect. 10.4.2). However, it does not mean that all of them play once at the same time. In each part musicians play alternately, depending on musical pieces.

There exist more than 300 professional orchestras in the world.

10.4.2   Types of Orchestras

There are various types of orchestras with different sizes. Traditionally they are classified according to the number of wind instruments played: single wind, double winds, triple winds, quadruple winds, and maybe larger. The numbers of the musicians for each part in the different size of the orchestra are listed

---

[4]A large cantata on poems by the Danish novelist Jens Peter Jacobsen. The "Gurre" refers to Gurre Castle in Denmark, scene of a medieval love-tragedy.

in Table 10.1. The numbers may differ, depending on orchestras as well as musical pieces played.

Table 10.1 Orchestration

Group	Instruments	Single wind	Double winds	Triple winds	Quadruple winds
Strings	1st violin	8	10	14	16
	2nd violin	6	8	12	14
	Viola	4	6	10	12
	Cello	2	4	8	10
	Double bass	2	4	6	8
Woodwinds	Piccolo	*	*	*	1
	Flute	1	2	3	3
	Oboe	1	2	2	3
	Engl. horn	*	*	*	1
	Clarinet	1	2	3	3
	Bass clarinet	*	*	*	1
	Bassoon	1	2	2	3
	Contrabassoon	*	*	1	1
Brass instruments	Horn	2	2–4	4–6	6–8
	Trumpet	1	2	3	4
	Trombone	1	2	3	4
	Tuba	*	1	1	1
Percussion	Timpani	*	1	1	1
	Others	*	2–3	2–3	2–3
Harp		*	1	1–2	1–2
Keyboard		*	1	1–2	1–2
Total		~ 30	~ 50	~ 75	~ 100

10.4.2.1 Chamber Orchestra

A chamber orchestra is a rather small ensemble of the single wind or double winds type. It means that the number of the players in each instrumental group is small. Sometimes percussion instruments, the tuba or the trombone are not included. Usually such orchestras are specialized on Baroque and classic music of Bach, Händel, and Mozart. The Court Orchestra in Mannheim (double winds) was the standard orchestra till the late 18th century.

10.4.2.2 Symphony Orchestra/Philharmonic Orchestra

A larger orchestra is called a symphony orchestra or philharmonic orchestra. However, there is no clear border between a chamber and a symphony orchestra. In the symphony orchestra each part is occupied by more than one person (double winds or higher winds). From Beethoven on a larger orchestra was required.

10.4.2.3   String Orchestra

A string orchestra is an orchestra consisting solely of string instruments such as the violin, the viola and the cello (occasionally the double bass is included) as shown in Fig. 10.9. Similar to the symphony or chamber orchestra, the violins are divided into first and second violins. The size differs from 12 to 60. Famous pieces are "Eine kleine Nachtmusik KV 525" (a small serenade) of Mozart and the Serenade for strings in C major Op. 48 of Tchaikovsky. Béla Bartok (1881–1945) composed "Divertimento" for string orchestra, Sz. 113, BB 118 in 1939. Edvard Grieg's "Holberg Suite Op. 40" was originally composed for the piano, but later rewritten by himself for string orchestra.

Fig. 10.9 Outlook on a string orchestra: The Munich Chamber Orchestra performing as a string orchestra at a concert in the modern Pinakothek, München (CC BY-SA 2.0 de by Andreas Fleischmann)

10.4.2.4   Wind Orchestra

A wind orchestra is also called a concert band, wind ensemble, symphonic band, wind symphony. It consists of the woodwind instruments, brass instruments, and percussion instruments. In other words, it is an orchestra without string instruments. However, occasionally the double bass and/or bass guitar are included for basso continuo. A military band belongs to the wind orchestras. Examples of pieces composed for the wind orchestra are: Music for the Royal Fireworks HWV 351 of Georg Friedrich Händel, Serenade No. 10 for winds in B♭ major KV 361/370a (also known as Gran Partita) of Wofgang Amadeus Mozart, and First Suite in E♭ for military band of Gustav Holst.

10.4.2.5   Big Band

A big band is a type of musical ensemble for jazz music that usually consists of ten or more musicians with four sections: saxophones, trumpets, trombones, and a rhythm section. Big bands originated during the early 1910s and dominated jazz in the early 1940s, when swing was most popular. Figure 10.10 shows the famous Whiteman band. The term "big band" is also used to describe a genre of music, although this was not the only style of music played by big bands.

Fig. 10.10 Whiteman band 1921 (public domain: photo from sheet music cover in the collection of Fredrik Tersmeden (Lund, Sweden))

10.4.3 Seating Arrangement

The seating arrangement of a modern orchestra is a sophisticated matter, in which the sound characteristics of instruments, the radiation direction of instruments and interaction between various groups of instruments have to be considered. The orchestra plays in a concert hall which is designed according to a hopefully optimal room acoustics (see Chap. 12). There have been many possibilities for the seating arrangement in history. Among them two competitive arrangements described below have survived.

- German arrangement (viewed from the audience): The 1st violins sit at the front left, the 2nd violins at the front right, double basses behind the 1st violins, violas left of the 2nd violins, cellos are in the middle. This arrangement was used till the beginning of the 20th century. A disadvantage is the difficulty for balancing the sound of the 1st and 2nd violins.
- American arrangement: As shown in Fig. 10.11, the 1st violins sit at the front left, the 2nd violins are right behind the first violins, cellos at the front right and violas left of the cellos, double basses behind the cellos and the violas. This arrangement is now a world-wide standard, except in cases of historical performances. Here each sound distribution can be better heard. Different sounds can merge or enter into dialogue with each other.

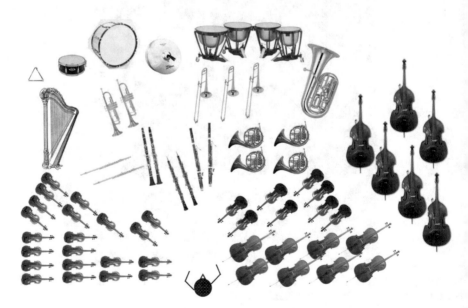

Fig. 10.11 The American seating arrangement of the orchestra

10.4.4   People

10.4.4.1   Conductor

In most of the modern orchestras there is one person who stands at the center of the front stage, back-facing the audience without playing any instrument. This person is the conductor who leads the performance of the orchestra.

A young conductor said to me one day that he is asked often by amateurs to teach them how to conduct. At first glance it looks very easy: You do not need to play any complicated instrument, but just stand in front of the orchestra and move your arms. However, even to stand in front of the orchestra "straight" and move down the arms "exactly perpendicularly" are already big problems for laymen.

So, a conductor is standing straight in front of the orchestra and starts moving his arms.... as shown in Fig. 10.12. Why?

Fig. 10.12 Conductor Ernst von Schuch moves down his arms exactly perpendicularly, concert master Ludwig van Beethoven is ready to play.... (illustration of the conductor: public domain by Robert Sterl, illustration of the concert master: M. Covarrubias)

It is almost impossible to expect that more than 50 people can be organized to do something together without any leader. A self-organization does not really work. Even if these people are reasonable and modest, this is very difficult. If they are scientists or musicians, things get worse and chaotic. We have no idea what to do with 50 scientists who are eager to talk about their own research.

But in the case of an orchestra, fortunately, a conductor is standing straight and tries to organize more than 50 musicians.

In a concert, at first and at least, musicians should know when the music starts. And then they have to know with which speed the piece should be played. For the cyber players these two factors (the starting time and the speed) are enough for playing together keeping in time and finishing together. The conductor is needed only for a few seconds at the beginning: a signal for start and tempo for a few bars. If the tempo should be changed, the conductor gives a corresponding signal.

In the real world, however, even these two factors are not so simple. Each instrument has a different response time: the time delay from the moment when the player starts playing to the moment when the tone is created [6]. Furthermore, some people can react quickly to the signal which the conductor gives, and others a little bit slowly. A good conductor can find some compromises among different instruments, as well as different human beings. And nevertheless, he knows when exactly the performance will finish by the second–very punctual like Japanese trains.

In an opera performance, if a singer stands far away from the orchestra and starts singing by hearing the orchestra, usually the song comes a little bit later than it should. There is a delay between the time when the orchestra plays and the time when the singer hears it. If the singer starts singing by looking at the signal given by the conductor, there is no such a delay. Therefore, it is important for the conductor to give a signal at the right moment. It is good that light travels much faster than sound. Once the singer starts singing, however, the conductor should lead the orchestra, so that the orchestra accompanies the singer appropriately.

To conduct an orchestra for a ballet performance is again different from an opera performance. If the orchestra varies the tempo or rhythm a priori, the ballet dancers get crazy, stumbling over their own feet. The important work for the conductor is, therefore, to keep the tempo and rhythm constant for a certain time. Furthermore, the conductor should consider the size of the stage. There is a tendency that dancers feel music faster in a smaller hall, slower in a larger hall, probably because of the sound reflections from the walls.

All mentioned above constitutes matters before music. The main task of a conductor is to present the music. He has to study the score and construct the performance in his own way. He has to know not only the written score but also the character of the individual composer. Each composer writes scores in his own way. Therefore, for example, a *ppp* of an Italian composer may be just a *p* for a German orchestra, and vice versa.

During a performance some accidents often happen. A string of a violinist is broken, a sheet of musical notes falls down, a percussionist is not awake when he should crash cymbals, and, and.... The conductor should keep calm and react spontaneously for the optimum solution without letting any audience notice it.

We can not write, how conductors study scores and construct their performance, because each conductor has his/her own way. What we can surely say is that to work as a conductor is a very hard job.

Intermezzo

Further, one more important task for a conductor is to establish a good relationship with the orchestra members. A conductor can be compared to an employed executive. Both should work as a supervisor of a group which they have not known before. They should not behave too arrogant, but they have to keep their own way. To obtain a good result they and the people of the group should understand each other to a certain extent. Since understanding individual persons is too time consuming, they start contacting section leaders. If they obtain support from leaders in rather difficult sections, then they are on the right way. If a conductor has an instinctive aversion to the concert master, the leader of percussion instruments or the oboe or flute players, this is fatal.

10.4.4.2   Concert Master and Co.

The concert master of an orchestra is a violinist who sits in the first row at the front side of the stage. Before the conductor shows up, he stands up and asks the oboe player to give the concert pitch A. He is the person with whom the conductor shakes hands at the end of the concert.

Once upon a time, when an orchestra consisted of about 20 persons, the concert master was the person who gave the signal to start and who gave signals for tempo and dynamics. Since the orchestra has become larger, a funny person who is called "conductor" appears without violin and takes the concert master's job away.

What does the concert master do except asking the oboe for the tone A and shaking hands with the conductor? Of course, he is a leader of the group of the 1st violins. He leads the 1st violins. Then, is he the same as other leaders in other parts like the 2nd violin, the viola...?

A concert master is a "host of the orchestra house": This is a word of Fuminori Shinozaki, the concert master of the NHK Philharmonic Orchestra [7]. He welcomes the guests (the conductor and soloists) on behalf of the orchestra house, and treats them so that they feel comfortable. If the guests feel fine and make a good job with the orchestra members, his duty is done well.

But sometimes the things get a little bit more difficult. He observes the members during performance. Usually he knows the orchestra members much better than a guest conductor. He sees that some members do not understand what the conductor means, and they look at him for help. Then, he functions as a shadow conductor.

Now look at the members of the orchestra house. They are musicians and specialized in one (or some) instruments. There could exist a close relationship between the instrument and its player. It is like the face of a dog owner looking alike his (her) dog. There are a lot of jokes about such a relationship. If we would start writing about these, we could not finish this book. Let us skip it.

10.5   Acoustics of an Orchestra

10.5.1   Directional Effects

When instruments form a musical ensemble, it is vital for the players to receive and notice the music emitted by the other instruments. For this, each individual member has to know the directional characteristics of his (her) own and the other instruments and make optimal arrangements according to these.

In the following two sections we explain basic directional characteristics from an elementary sound source and a linear radiation group [8].

10.5.1.1   Directional Effects of Elementary Sound Sources

When the form of an acoustic wave has a spherical shape, the sound radiates isotropically in all direction. This is a radiation of the 0th order, emitted by a "monopole". In this case the direction characteristics is a sphere. In the plane the direction diagram is a circle with the source in its center.

When such a spherical sound source oscillates between one point and another, the radiation is of first order, or a "dipole". A sound source of first order can be mathematically interpreted as two 0th order sources at distance d, which oscillate with a phase difference π and the same amplitude (see Fig. 10.13).

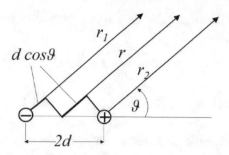

Fig. 10.13 For derivation of direction diagram of a first order radiation

Let us assume that the distance between the sound source at the center and the point of the receptor r is much larger than d ($r \gg d$). Then, the distances r_1 and r_2 are:

$$r_1 \approx r + d\cos\vartheta,$$
$$r_2 \approx r - d\cos\vartheta, \tag{10.1}$$

where ϑ is the angle between the dipole and observer.

The two spherical sound sources create the velocity potentials Φ_1 and Φ_2 at the point of the receptor (cf. Sect. 6.2.5.3):

$$\Phi_1 = -\frac{e^{-ikr_1}}{r_1} = -\frac{e^{-ik(r+d\cos\vartheta)}}{r + d\cos\vartheta},$$

$$\Phi_2 = +\frac{e^{-ikr_2}}{r_2} = +\frac{e^{-ik(r-d\cos\vartheta)}}{r - d\cos\vartheta}. \tag{10.2}$$

The total potential Φ is then

$$\Phi = \Phi_1 + \Phi_2 = \frac{e^{-ik(r-d\cos\vartheta)}}{r - d\cos\vartheta} - \frac{e^{-ik(r+d\cos\vartheta)}}{r + d\cos\vartheta}. \tag{10.3}$$

Since the distance between the sound source at the center and the point of the receptor r is much larger than d ($r \gg d$), practically both waves have the same amplitude, and therefore, comparing to r, $d\cos\vartheta$ in the denominators can be neglected:

$$\Phi = \frac{1}{r}e^{-ikr}(e^{ikd\cos\vartheta} - e^{-ikd\cos\vartheta}) \approx \frac{e^{-ikr}}{r}sin(kd\cos\vartheta). \tag{10.4}$$

When the distance between the two spherical sound sources is very small ($kd \ll 1$)

$$\Phi \approx \frac{e^{-ikr}}{r} kd \cos\vartheta. \tag{10.5}$$

The velocity potential Φ contains the factor kd, which is proportional to the frequency.

Furthermore, such a dipole radiation depends on direction. The maximum sound pressure appears in the direction along the connection line of two monopoles ($\vartheta = 0$ or π). In the symmetry plane ($\vartheta = \pi/2$) the two radiated waves cancel each other because of interference. With Γ called the radiation direction factor, the direction diagram ($\Gamma = cos\vartheta$) of the dipole radiation source has thus the shape of the numeral 8, as illustrated in Fig. 10.14.

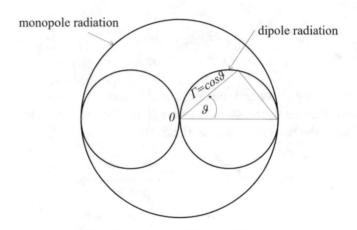

monopole radiation

dipole radiation

Fig. 10.14 Direction diagram of a zero and a first order radiation

The power of the first order sound source at the lower frequency is a factor $(kd)^2$ smaller than in the case that one of the two spherical sources would be turned on. The weak radiation is plausible: The high pressure which is created from one spherical source is compensated by the low pressure of the neighboring sphere.

The sound radiation of the flute can be explained by this dipole radiation.

10.5.1.2   Directional Effects of a Linear Radiation Group

Here we explain an example, which is a very simple but important radiation group: Many spherical sound sources (with number n) line up on a straight line with a constant mutual distance d as shown in Fig. 10.15.

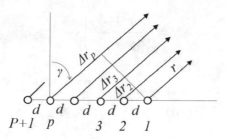

Fig. 10.15 For derivation of direction diagram of a linear group of n sources, the mutual distance of which is constant d

The angle between the line of the sound sources with the observing direction is γ. The distances between each sound source and the point of the receptor are different. The distance for the pth is shorter than that for the first source by:

$$\Delta r_p = (p-1)d \cdot sin\gamma. \qquad (10.6)$$

The amount of the sound pressure is the sum of the sound pressures which are created by each elementary source. The relative amount can be described as a radiation direction factor Γ which is a function of angles φ and ϑ:

$$\Gamma = \left| \sum_{p=1}^{n} \frac{1}{n} e^{-ik(r+\Delta r_p)} \right| = \frac{1}{n} \left| \sum_{p=1}^{n} e^{-ik(p-1)d \cdot sin\gamma} \right|. \qquad (10.7)$$

Finally,

$$\Gamma = \frac{1}{n} \left| \frac{e^{-iknd \cdot sin\gamma} - 1}{e^{ikd \cdot sin\gamma} - 1} \right| = \frac{1}{n} \left| \frac{sin(n\varphi)}{sin\varphi} \right|, \qquad (10.8)$$

where

$$\frac{kd}{2} sin\gamma = \frac{\pi d}{\lambda} sin\gamma = \varphi.$$

In order to present the direction characteristics of the linear group we use the function:

$$y = \left| \frac{sin(n\varphi)}{n \cdot sin\varphi} \right|. \tag{10.9}$$

When $\varphi = 0, \pi, 2\pi, \ldots$, the function y approaches the maximum value 1

$$y = \left| \frac{n\varphi + \cdots}{n\varphi + \cdots} \right| \to 1. \tag{10.10}$$

Note that for $x \to 0$:

$$sin(nx) \approx nx, \quad sinx \approx x, \ldots$$

This means that the direction factor Γ has maxima at $\varphi = 0, \pi, 2\pi, \ldots$. The function y disappears between $\varphi = 0$ and $\varphi = \pi$ for the (n-1)th angle:

$$\varphi = \frac{\pi}{n}, \frac{2\pi}{n}, \ldots, \frac{(n-1)\pi}{n}.$$

Between two "0" positions there exists one "submaximum", that is, between two maxima there exist (n-2) submaxima.

If $dy/d\varphi = 0$,

$$n \cdot tan\varphi_m = tan(n\varphi_m), \tag{10.11}$$

where φ_m is the angle for the maxima.

Then,

$$y_m^2 = \frac{1}{n^2 sin^2\varphi_m + cos^2\varphi_m}. \tag{10.12}$$

This is the equation for an ellipse with main axes 1 and $1/n$, which is the geometrical location of the maximum direction characteristics of the linear sound sources.

Figure 10.16 shows the function $y = |sin(n\varphi)/n \cdot sin\varphi|$ for the case of $n = 6$ and $d = \lambda/2$. Such types of direction diagram are seen in the acoustic radiation of many instruments (see Sect. 10.5.2).

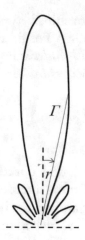

Fig. 10.16 Direction diagram of a linear radiation group for the case of $n = 6$ and $d = \lambda/2$

10.5.2 Acoustic Radiation of Instruments

Each instrument has a specific radiation direction of the sound, depending on its shape and frequency.

Common among various instruments is a spherical radiation at low frequency, where the wavelength is large in comparison to the sound source. This is a radiation of the 0th order. Such wavelengths correspond to fundamentals of the instruments. They are, for example, 200–400 Hz for a violin, 300–350 Hz for a flute, 60–100 Hz for a horn, 30–80 Hz for a tuba, and so on. Above 500 Hz such a spherical radiation is observed less [9].

The radiation characteristics of each instrument varies, when more than two instruments are played closely together. Even for the simplest case of the spherical radiation of two instruments the characteristics of the sound emission is then changed from the monopole emission to the dipole emission. More generally, not only the distance between two instruments, but also frequency, involvement of harmonics, phase and strength affect the resulting direction characteristics as a group.

The seating arrangement of the orchestra (see Sect. 10.4.3) must have been introduced from experiences and experiments of such radiation characteristics of each instrument, as well as many instruments as a group. Furthermore, reflections from walls or the ceiling should be also considered, as discussed in Chap. 12.

Below we explain some relationships between the acoustic radiation of instruments and the orchestra seating. For details we recommend the book of Jürgen Meyer's work on *Acoustics and the Performance of Music: Manual for Acousticians, Audio Engineers, Musicians, Architects and Musical Instruments Makers* [9].

10.5.2.1 Brass instruments

Brass instruments radiate their sound directly from the bell opening. The characteristics of the sound radiation direction of these instruments (except horns) is rotationally symmetric around the bell axis. Therefore, it is affected by the shape and size of the bell, as well as the connected conical part of the bore. Since one has almost no vibration of the metal tube, there is no influence of the material. In principle, the radiation angle for a trumpet, trombone and tube becomes narrower with increasing frequency. However, the frequency dependence of the angle shows maxima at the formants frequencies (about the formant, see Sect. 8.3). Figure 10.17 illustrates the wavelength dependence of the radiation angle of a trumpet for 0–3 dB [9].

Trombones and tubas show a similar tendency: Just the wavelength shifts to the lower frequency.

On the other hand, horns do not show the rotational symmetry in their sound direction. The direction characteristics of the sound is very complicated, because

- the shape is far from a rod shape, but almost like a disc,
- the horn player puts his right hand into the bell opening during playing,
- the sound is bent around the player's body, since the horn is held close to the player.

The tones from 150 to 500 Hz in the horizontal plane are directed toward the right of the player forming a semicircle. At higher frequencies the radiation angle is narrower. There is a general preference of the right side and the back side [9].

Figure 10.18 shows a main radiation directions of the sound and seating arrangement for the brass instruments in an orchestra.

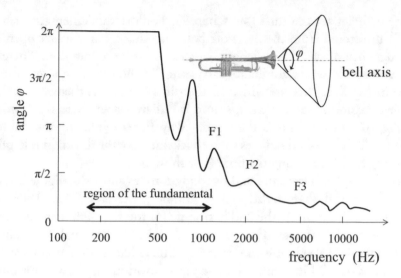

Fig. 10.17 Wavelength dependence of the radiation angle ϕ of trumpets for 0–3 dB; F1, F2, F3: formant regions [9]

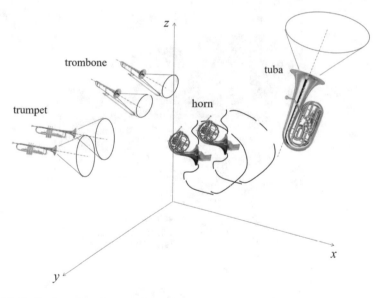

Fig. 10.18 Radiation directions and seating arrangement of brass instruments, x: direction to the audience, xy-plane: floor, z: direction to the ceiling

10.5.2.2 Woodwind Instruments

Among woodwind instruments flutes and piccolo are instruments without reed. The sound of the flute (and piccolo) is emitted from the blow-off hole and open holes on the tube (see Chap. 7). For the fundamental tones (C4 to D5) the sound from the blow-off hole and that from the first open hole has a distance of half a wavelength, but the phases are identical. Therefore, these two sound waves annihilate each other in the direction along the axis of the flute, while they are added in the direction perpendicular to the axis (dipole emission). Consequently, sound is radiated strongly to the front side (see Fig. 10.19 top). This is the reason, why flutes are facing the public in concert halls. The sound radiates to the backside, too: the direction is in mirror symmetry to the line along the flute (not shown).

Also the higher harmonics and overblown tones show the dipole behavior in the directional characteristics. The middle panel of Fig. 10.19 illustrates the radiation direction for the 2nd harmonic (or the 1st overblown harmonic), where the length of the standing wave is half of the fundamental. The radiation to the backside is also in mirror symmetry. The radiation direction for the 3rd harmonic is divided into three directions as shown in Fig. 10.19, bottom. To the backside only the sound which corresponds to the middle direction is emitted (not shown).

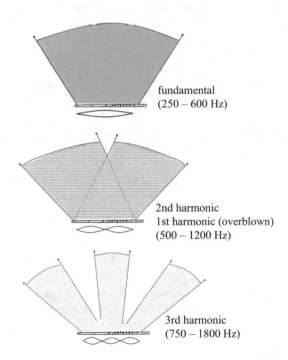

fundamental
(250 – 600 Hz)

2nd harmonic
1st harmonic (overblown)
(500 – 1200 Hz)

3rd harmonic
(750 – 1800 Hz)

Fig. 10.19 Principal radiation directions towards the front of flutes in the horizontal plane for fundamental (top), 2nd harmonic and 1st overblown harmonic (middle) and 3rd harmonic (bottom). In each panel corresponding standing waves are shown

Oboes are reed instruments. Different from the mouth hole of the flute, the reed does not contribute to the sound radiation. Therefore, oboes do not create dipole radiation. In the case of oboes sound is radiated not only from the first open hole but also from all open flaps. The strongest radiation is directed to the front around 1000 Hz. The audience hears the sound directly and also the reflection from the ceiling up to 2000 Hz. At higher frequencies reflection from the floor dominates. Within the orchestra these tonal contributions are strongly absorbed by the musicians sitting in front, so that the highest tonal component is weakened. This is, however, comfortable for the audience, since the high tone of the oboe is often perceived as shrill.

The radiation direction of clarinets is similar to that of oboes. Since the bell of clarinets is larger, the radiation angle is wider than that of oboes.

Different from other wind instruments the bassoon is upwardly directed. Therefore, the reflection from the ceiling should be well considered. Around 250 Hz the intensity of radiation is almost spherical. For the higher frequencies the sound is radiated to a direction with a narrower angle.

Usually the wood wind instruments are placed directly behind the string instruments and in front of some brass instruments. Figure 10.20 gives some ideas how their sounds are emitted.

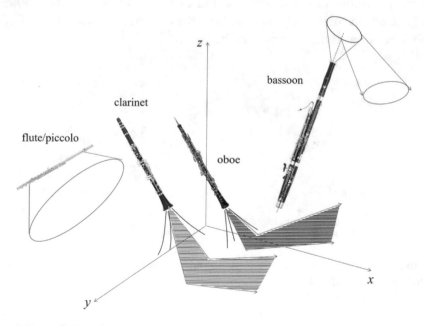

Fig. 10.20 Radiation directions and seating arrangement of wood wind instruments, x: direction to the audience, $x - y$ plane: floor, z: direction to the ceiling

10.5.2.3 String instruments

The sound of string instruments emanates mainly from the resonance body of the vibrating wood. Sound of low frequency is emitted also through the f-hole. Moreover, each instrument has some different properties for the radiation characteristics.

The principal radiation direction of violins between 200 and 500 Hz is spherical. At higher frequencies both the vertical and horizontal radiation angles vary. However, sound is radiated in the angular region shown in Fig. 10.21 at any frequency. The high frequencies which can not be played are heard as harmonic series (2000–5000 Hz).

The seating of the first violins in an orchestra optimizes the sound radiation toward the audience for the whole frequency region. Especially in the frequency region between 1000 and 1250 Hz, where the radiation angle is narrow, a strong, bright tone color from the first violins can be heard. For the American seating arrangement the radiation direction of the second violins is not so different from that of the first violins. However, the sound of the second violins is hidden by the first violins. On the contrary, the sound in this range from the second violins in the German seating arrangement is poor. However, the second violins can radiate the sound of the frequency around 1500 Hz much better than the first violins, because the horizontal radiation at 1500 Hz is directed towards the left from the bridge.

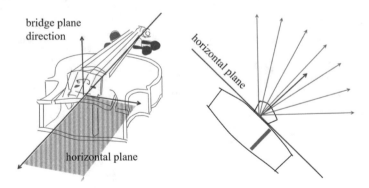

Fig. 10.21 Left: direction of the bridge plane and horizontal plane, right: angular region of violins at any frequencies

Due to their shape and size the manner of sound radiation of violas is similar to that of violins. However, since the sound emission through the f-hole is more profound, and since the individual radiation characteristics of violas varies more than that of violins, it is not so easy to locate the viola in an orchestra. On the average, the components between 600 and 700 Hz are radiated to the audience in the American arrangement, resulting in a darker color.

Coffee Break

Although violas create a very comfortable sound, they belong often to the most unfortunate instruments. They are squeezed between the 2nd violins and cellos, and many people do not notice whether ever any sound comes from them.

Correspondingly, there are hundreds of "viola jokes", some of which are very nasty and insult viola players.

****

A viola player always opens and closes his locker before he goes to the stage. Other members of the orchestra wonder why. One day a colleague opened his locker, and found a piece of paper sticked on the rear side of the door. On the paper it is written: viola left, bow right!

****

Such a joke is of course not fair for viola players. In comparison with the 1st violinists they are very modest and, maybe therefore, often one of them is selected to the board of the orchestra.

Concerning cellos, only the lowest frequency C2 (65 Hz) radiates in circular form. For the frequency range 350–500 Hz the sound is radiated prominently toward the front. The reflections from the ceiling as well as the floor are effective. From 800 Hz on, the audience does not hear the direct sound, but the reflections from the ceiling and hanging. At much higher frequency (up to 2000 Hz) the sound is reflected a lot from the floor, but it is often hidden by other players.

The sound radiation for a double bass also depends on the frequency. However, it is directed mainly towards the front. A preferred region of radiation is the front semi-circle around the bridge.

Figure 10.22 shows some ideas of the sound direction for string instruments in an orchestra.

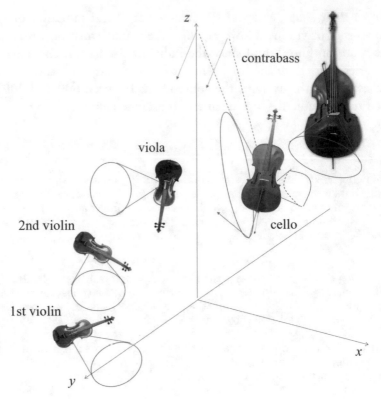

Fig. 10.22 Radiation directions and seating arrangement of string instruments, x: direction to the audience, xy-plane: floor, z: direction to the ceiling

10.5.2.4 Percussion instruments

In an orchestra percussion instruments are set at the back side.

Sound of a timpani is emitted upward from the surface of the membrane. The shape of the radiated sound field is, therefore, determined by the modes. With the first ring mode (01) the entire membrane vibrates in phase, resulting in a spherical characteristics. However, the main tone is generated from the mode (11) and additionally its harmonics. The mode (11) is the first radial mode in which the membrane vibrates in two half regions in opposite phase, resulting in rotationally symmetric characteristics around the horizontal axis (dipole). The mode (21) shows a quadrapole character and the sound direction has a clover-like form. For details on modes, see Sect. 8.2.1.

In spite of their low pitch, the timpani are easily localized by the audience. Therefore, the location of the timpani should be well considered for making cooperative sound with other instruments. Usually the timpani are put in the middle of the back side of the orchestra in order to obtain tonal symmetry.

Note that there are two different set-ups: the German style and the American style. In the German style the highest-pitch timpani is put at the left, and the lowest pitch timpani at the right. In the American style this is the other way around.

Different from timpani, the sound of drums radiates upward, as well as towards the bottom. For the small drums (like a snare drum) the preferable vibration mode is that both the upper and lower membranes vibrate in parallel (in opposite directions). Such radiation has a dipole characteristics [10]. When the drum has only an upper membrane, the drum shows the dipole characteristics. If the lower membrane is mistuned, it does not vibrate so much, resulting in a uniform radiation in all directions.

Large drums have a relatively uniform radiation [11].

Figure 10.23 shows schematically the radiation of a timpani, a snare drum and a bass drum in an orchestra.

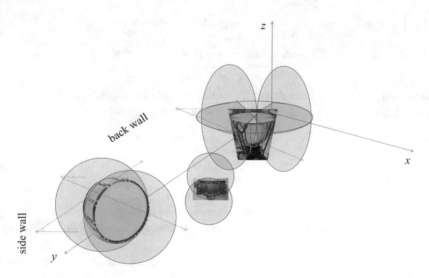

Fig. 10.23 Radiation directions and seating arrangement of timpani and drums, x: direction to the audience, xy-plane: floor, z: direction to ceiling

Coffee Break

Well done, Papa!

The timpanist enjoys to play the second movement of The Symphony No. 94 of Joseph Haydn "The suprise", better to say in German original "Mit dem Paukenschlag" (with the timpani beat).

tutti: timpani strong!

Bars 1 to 16 of the first violin part in the second movement. The "surprise" arrives in measure 16, when the rest of the orchestra joins the first violins in a fortissimo G-major chord

Haydn suggested extra that the timpanist should hit the timpani as strong as possible, so that sleeping audience would be awakened.

Acknowledgments Authors thank Motonori Kobayashi, a conductor of the Dortmund Philharmonic Orchestra for talking about being a conductor.

References

1. C. Ehrlich, *The Piano: A History* (Clarendon Press, Oxford, 1990)
2. J.H. Baron, *Intimate Music : A History of the Idea of Chamber Music* (Pendragon Press, Sheffield, 2003)
3. H. Heimeran, B. Aulich, *Das stillvergnügte Streichquartett* (Ernst Heimeran Verlag, München, 1936)
4. J. Spitzer, N. Zaslaw, *The Birth of the Orchestra: History of an Institution, 1650–1815* (Oxford University Press, Oxford, 2005)
5. R. Wagner, *Über das Dirigieren* (Insel-Verlag, Leipzig, 1869)
6. A. Farina, A. Langhoff, L. Tronchin, Realisation of "virtual" musical instruments: measurements of the impulse response of violins using MLS technique, in *Proceedings of CIARM95, 2nd International Conference on Acoustics and Musical Research, Ferrara*. Accessed 19–21 May 1995

7. K. Kondo, *Orchestra World* (Yamaha Music Enertainment, Tokyo, 2019), p. 125
8. E. Meyer, E.-G. Neumann, *Physikalische und Technische Akustik* (Friedrich Vieweg und Sohn, Braunschweig, 1974), pp. 135–145
9. J. Meyer, *Acoustics and the Performance of Music*, 5th edn. (Springer, Heidelberg, 2009), pp. 263–343
10. T.D. Rossing, I. Bork, H. Zao, D.O. Fystrom, Acoustics of snare drums. J. Acoust. Soc. Am. **92**, 84 (1992)
11. H.F. Olson, *Music, Physics and Engineering* (Dover Publicatrions, New York, 1967)

Part V
Hearing

Two old men are sitting together and hearing music:

A says: Is it Mozart?
B says: No, it is Mozart!
A says: Oh, I thought it is Mozart......

—modified from a Japanese joke

11

Physiology and Psychoacoustics

11.1 Hearing—Physiological Aspects

Certainly there are a lot of books on human biology and acoustics in which details about hearing are described. In this chapter we intend to explain the basic characteristics of hearing without using too many medical/anatomical terms.

11.1.1 Structure of the Human Ear

Figure 11.1 illustrates the anatomical structure of the human ear, which can be divided into three parts, the outer ear, the middle ear and the inner ear.

11.1.1.1 Outer Ear

The outer ear consists of the auricle and the ear canal. The auricle is the visible part of the ear that resides outside the head. It has roles not only for capturing sound but also for determining the incident direction of sound. The asymmetrical and uneven surface of the auricle functions as an acoustic resonator exited at the sound input from a certain direction, so that we can localize where the sound comes from. The ear canal terminates at the eardrum (tympanic membrane).

© Springer Nature Switzerland AG 2021
K. Tsuji et al., *Physics and Music*,
https://doi.org/10.1007/978-3-030-68676-5_11

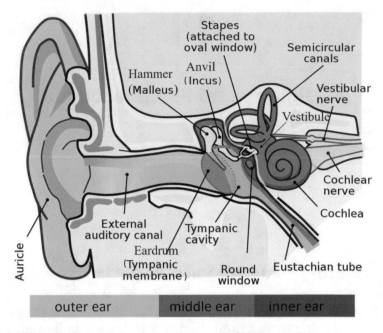

Fig. 11.1 Structure of the ear (CC BY 2.5 by Lars Chittka and Axel Brockmann [1])

11.1.1.2 Middle Ear

The middle ear is a region between the eardrum and the oval window. Behind the eardrum there are three small bones (ossicles): the hammer (malleus), the anvil (incus) and the stapes ending at the oval window (see Fig. 11.2). In this area an impedance conversion is taking place mechanically.

The hollow space of the middle ear is called the tympanic cavity. The round window connects the middle ear and the tympanic canal of the inner ear. The auditory tube (eustachian tube) links the middle ear with the nasal cavity (nasopharynx), allowing pressure to equalize between the middle ear and the throat.

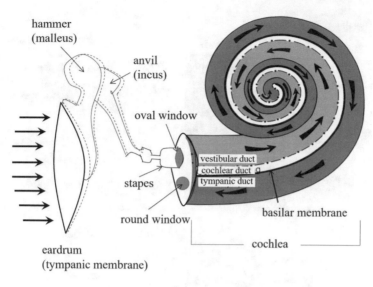

hammer
(malleus)

anvil
(incus)

oval window

vestibular duct
cochlear duct
tympanic duct

stapes

basilar membrane

round window

cochlea

eardrum
(tympanic membrane)

Fig. 11.2 Schematic presentation of sound transmitted through eardrum, hammer, anvil and stapes on the oval window of the inner ear, and the cochlea

11.1.1.3 Inner Ear

In the inner ear there is the bony labyrinth (the rigid, bony outer wall), which consists of three parts: the vestibule, semicircular canals, and the cochlea (see Fig. 11.1). The vestibule is the central part of the bony labyrinth and is situated medial to the eardrum. The semicircular canals are important for the sense of balance, and the cochlea is responsible for our hearing.

The cochlea is a spiral-shaped cavity as shown in Fig. 11.2. The name cochlea derives from Ancient Greek κοχλζας (kokhlias), meaning spiral or snail. It is filled with two liquid systems (perilymph in the vestibular duct and the tympanic duct, and endolymph in the cochlear duct). The basilar membrane and hair cells, as well as the tectorial membrane are located in the cochlear duct and this core part is called the "organ of Corti" (see Fig. 11.3).

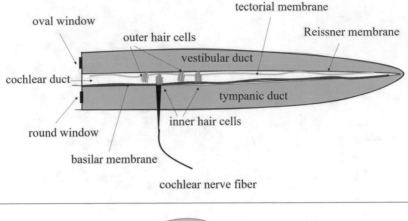

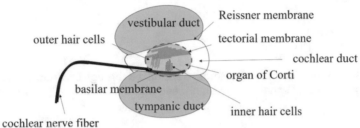

Fig. 11.3 Top: longitudinal section, bottom: cross section of the cochlea

11.1.2 Function of the Human Ear

The path of the sound is:
auricle → ear canal → eardrum → ossicles → vestibule → cochlea → cochlea
nerve → brain.
Sound pressure waves are funneled through the ear canal of the external ear to
the eardrum, and mechanically transduced by the bones (ossicles) in the middle
ear into a fluid-filled chamber, the vestibule, and further into the cochlea.
(There is also the phenomenon of bone conduction, which is the conduction
of sound to the inner ear through the bones of the scull.[1] This way people hear
their own voice.) The **basilar membrane** (details see below) in the cochlea hosts
sensory hair cells and they convert the mechanical waves to electric impulses.
These electric signals are passed through the cochlea nerve to the brain.

Humans can perceive sounds only in a limited frequency range and at limited levels of sound pressures. The area between the acoustic threshold and
(acoustic) pain threshold is called the auditory sensation area, which lies be-

---

[1] Ludwig van Beethoven is said to have used bone conduction after losing his hearing by placing one end
of a rod in his mouth and resting the other end on the rim of his piano [2].

tween 20 and 20,000 Hz. Both the sensitivity and noise tolerance of the ear are important. The lowest sound pressure perceived by humans is about $20\,\mu\text{Pa} = 2\cdot10^{-5}$ Pa, corresponding to $L_p = 0$ dB at 2000 Hz (see Table 6.1 in Chap. 6).

This pressure corresponds to a displacement of air molecules of only 0.1 Å, which is smaller than the mean free path of Brownian motion. The sound pressure by the thermal motion of air amounts to 1×10^{-6} Pa. Thus, nature has increased the ear sensibility just so far that it does not perceive thermal motion.

The sound pressure change Δp is transferred through the eardrum and ossicles in the middle ear into the inner ear, and finally the auditory impression is constructed in the ear-brain system. It takes less than 0.1 s for the ear to transform an acoustic signal to a nerve impulse. A sensory impression is created by an energy of 10^{-18} J (very high sensitivity). The pain threshold is 130 dB, which is a more than 3×10^{6} times higher sound pressure than the lowest perceivable one.

11.1.2.1 Basilar Membrane

The basilar membrane is a membrane-like textile in the cochlear duct as shown in Fig. 11.3. On the basilar membrane the hair cells (receptors for auditory perception) are attached.

When the stapes press on the oval window, this causes the liquid of the vestibular duct to move. The oval window has only approximately 1/18 the area of the eardrum, therefore, it produces a much higher pressure. The liquid motion in the cochlear tube and subsequently the motion of the basilar membrane [3] generate traveling waves[2] along the tube [4]. These waves start at the oval window and reach the round window through the vestibular duct, the very narrow turning point at the apex of the cochlea and the tympanic duct. Accordingly, the fluid in the cochlear duct moves, too. Since these liquids are almost incompressible, the round window (an elastic membrane) moves out to compensate the pressure.

It is known that high frequencies cause a peak wave near the base of the cochlea, and low frequencies produce their peaks toward the apex.[3]

---

[2]Traveling waves are periodic waves propagating in a given spatial direction.

[3]The basilar membrane is a membrane with different width, stiffness, mass, damping, and duct dimensions at different points along its length. The properties of the membrane at a given point along its length determine its characteristic frequency, a frequency for which it is most sensitive. The basilar membrane is widest (0.42–0.65 mm) and least stiff at the apex of the cochlea, and narrowest (0.08–0.16 mm) and stiffest at the base (near the round and oval windows) [5].

Such a dependency of the frequency response is related to longitudinal membrane properties connected with hydrodynamic flux in the ducts. The nonlinearity of the vibration of the basilar membrane and the high-pass filtering effect may cause high selectivity of cochlear frequency [6].

11.1.2.2 Organ of Corti

The organ of Corti is the sensory organ of hearing, discovered by A. G. G. Corti in 1851. It is located in the cochlear duct and contains hair cells which are attached to the basilar membrane.

There are two types of hair cells, the outer and the inner hair cells. From the top of the hair cells tiny finger-like projections (stereocilia) protrude. Through mechanotransduction the stereocilia detect motion in their environment caused by a movement of the basilar membrane. The stereocilia form bundles of 30–300 and are arranged such that the shortest stereocilia are at the outside and the longest in the center. This gradation may play an important role for tuning capability.

Outer Hair Cells

The outer hair cells function as an amplifier of the low level vibrations which enter the cochlea. When the outer hair cells are depolarized,[4] the motor protein (prestin [7]) in the outer hair cells shortens quickly and subsequently pulls the basilar membrane, as well as the tectorial membrane. As a result the basilar membrane is more deflected and causes a more intense effect on the inner hair cells.

When cells are hyperpolarized,[5] prestin lengthens and eases tension of the membranes, so that the electrical signals caused by the inner hair cells become weaker. In this way, the hair cells protect the brain against too strong signals.

Inner Hair Cells

When the basilar membrane is in motion, it presses on the hair cells. The deflection of the stereocilia opens mechanically gated ion channels, so that any small cations enter the cell. Subsequently, the cells are depolarized and cause a receptor potential which opens voltage-gated calcium channels [8]. Calcium ions enter the cells and trigger the release of neurotransmitters at the basal end

---

[4]Depolarization is a change within a cell, during which the cell undergoes a shift in electric charge distribution, resulting in less negative charge inside the cell.

[5]Hyperpolarization is a change in a cell's membrane potential that makes it more negative. It is often caused by efflux of K^+ through K^+ channels, or influx of Cl^- through Cl^- channels.

of the cells. An electrical signal is then sent through the auditory nerve and into the auditory cortex of the brain as a neural message. This is a very important process of transformation of the sound vibration into electrical signals.

11.1.3 Tonotopic Mapping

The electrical signals from the inner hair cells propagate through the auditory nerves to the cortex. There are a lot of steps and processing of auditory information to reach the cortex. However, we concentrate on a few interesting topics for our theme "Physics and Music".

Tonotopy is the spatial arrangement in the brain where sounds of different frequency are processed. The original idea of tonotopy came from Békésy's work showing that different sound frequencies cause maximum wave amplitudes to occur at different places along the basilar membrane [4].

Each cochlear nerve connected to the different locations of the inner hair cells on the basilar membrane possesses the corresponding frequency information. The information is represented by the rate of action potentials as well as the particular timing of individual action potentials.

Such information from each cochlear nerve is projected one to one on the space in the primary auditory cortex through the cochlear nerve and associated midbrain structures [9]. The projection in the primary auditory cortex is called a tonographic map.

Not only the pitch but other characteristics such as sound intensity, tuning bandwidth or modulation rate may form similar maps in the cortex.

11.2 Hearing Music—Psychoacoustical Aspects

11.2.1 How Do We Hear Music?

How sound waves reach the primary auditory cortex has been briefly explained above. Now we try to explain how we recognize music from the information reaching our brain. There are many common aspects in language and music for hearing and understanding. We focus our explanations here on understanding melody, rhythm and harmony, listening to music and playing music.

We have a tonotopic map for pitch and other brain mechanisms coded in our cortex, for sound intensity, tuning bandwidth and other parameters. These data in isolation, however, do not mean so much. For example, even an alarm signal indicating danger composed of only a single pitch will not be recognized as such without additional features like its loudness and temporal dynamics.

To combine different kinds of information our brain has a special place, named the association cortex. The temporal association area is responsible for combining the variety of primary auditory information. The complex data are transferred to the prefrontal cortex, where the data are integrated over time.[6] And finally, we hear "music". This process is illustrated in Fig. 11.4. The information to the frontal lobe flows also to the primary motor cortex.

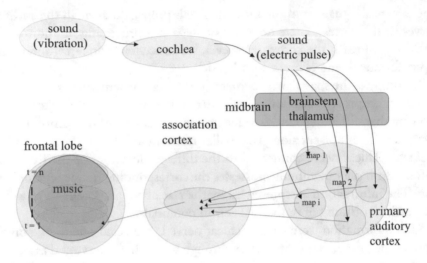

Fig. 11.4 Schematic presentation of our music hearing

11.2.2 A Huge Black Box

Is it clear now for you how we hear music, after reading the explanations above? We have to admit that the answer will be "No, it is not clear". There are many things, which are not yet clarified.

Let us assume that we hear somebody hitting the piano key of C4. The note has a pitch of 262 Hz. And the piano emits the periodical wave form shown in Fig. 11.5 top. In this case the period is $T = 1/262$ s. Our eardrum receives waves with this shape and vibrates accordingly. The vibrations are transmitted through the ossicles to the oval window, and then the basilar membrane and hair cells react to the vibrations and map the corresponding frequency.

At this point we stumble already. As described in Chap. 6 wave forms can be mathematically/physically transformed into a spectrum (see Fig. 11.5 bottom)

---

[6]It is true that we still lack a mechanistic explanation of how e.g. melodies are reconstructed across time. It is certainly not a simple discrete stacking.

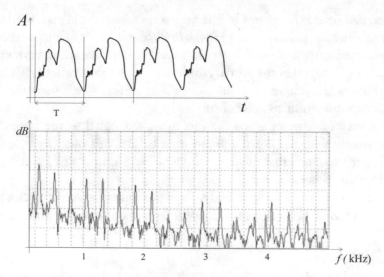

Fig. 11.5 Top: wave form of piano (C4), bottom: spectrum of piano sound (data taken from [10])

on the basis of the Fourier transform. The spectrum consists of a fundamental and many harmonics. Does the basilar membrane know the Fourier transform? In other words, are the reaction of the basilar membrane and the hair cells a kind of Fourier transform? If so, how does it function? We know that the sound pressure causes the liquid motion in the cochlear tube and subsequently the motion of the basilar membrane [3]. And we know also that high frequencies cause a peak wave on the basilar membrane near the base of the cochlea and low frequencies produce their peaks towards the apex. But how is the wave form actually transformed to the spectrum?

Another question arises: If the same tone is played also with a violin at the same time, we receive two different wave forms (a typical wave form of a violin is shown in Fig. 6.16). With closed eyes we recognize immediately that a violin and a piano are played together. Maybe our ear can distinguish these two tones because of the different directionality. But both enter into our ear and cause vibrations. How does the basilar membrane react separately to two different tones (with the same pitch)? Does the basilar membrane react to all vibrations without distinguishing two different sound groups and only later, in another place in our brain, they are separated?

Let's go further. So, the piano sound is transformed to the corresponding piano spectrum. Pitches are mapped on the basilar membrane, and this map is transcribed into the primary auditory cortex by electric pulses. Here comes the next question. How long is this map kept? If it is overwritten, in which time interval does it happen? Or is it only a transitional state for sending the signals to the association cortex? How are different kinds of maps (the pitch map, loudness map, and others) synchronized in the association cortex?

We simply accept that the primary auditory cortex sends various signals with appropriate synchronization to the association cortex. The association cortex reconstructs the sound complex. But how?

The association cortex sends the reconstructed signal to the frontal lobe. For the person who has heard a piano tone before, this memory is activated and this person knows that the tone which he hears now is also a piano sound. How does this process work?

Let us go even further. Somebody plays the following notes with a piano, and the tempo is 120 (120 fourth notes in a minute = 2 fourth notes per second).

Twinkle twinkle little star???

The pitch sequence is 525 Hz, 525 Hz, 784 Hz, 784 Hz, 880 Hz, 880 Hz, 784 Hz with a duration of 0.5 s for each tone. Then a rest for 0.5 s is followed by tones 698 Hz, 698 Hz, 659 Hz, 659 Hz, 587 Hz, 587 Hz, again with a duration of 0.5 s for each tone and terminated with 525 Hz after 2 s. Most of you can neither say that the frequency of the first tone is 525 Hz nor that it is C5. But you can say the pitch of the second tone is the same as that of the first one (perfect unison), that of the third one is the perfect fourth, and so on. The time course of the pitch changes is constructed in the frontal lobe. If one has heard "Twinkle twinkle little star" before, one recognizes that the melody heard is a well known melody, namely the beginning of the song "Twinkle twinkle little star".

Unfortunately, we have to admit that this process is also not well understood.

So we are standing in a huge black box, and wonder how beautifully our auditory system works. To study all our naive questions is beyond the scope of this book. Honestly speaking, the authors have rarely thought about how hard and how precise our ears and other auditory brain areas work all the time. After knowing this, we like to caress our ears and say "Thank you for the nice job, brave ears"! Let us take care of our fine auditory organs and enjoy music.

11.2.3 We Hear (or Do Not Hear) in Somewhat Different Ways…

What we described above with a huge black box is still just a standard physiological process, however. There are other complications which happen for everybody. We will introduce some examples. Figure 11.6 presents three cases where the real sound waves in the air are differently perceived. (For the standard case where one perceive sounds which exist in the environment, as shown in Fig. 11.6a.) Our examples are masked tones (b) [11], additional tones (c) and the cocktail party effect (d).

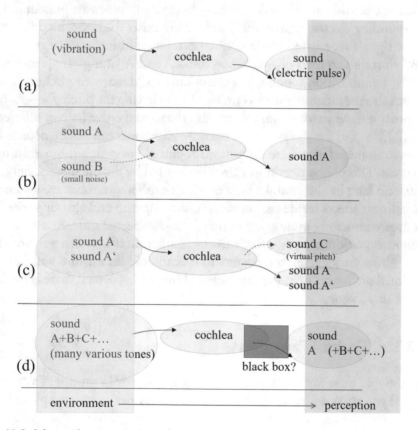

Fig. 11.6 Schematic presentation of sound waves in our environment and our perception: **a** normal case, **b** masked tone, **c** additional tone, **d** cocktail party effect

11.2.3.1   Masked Tones

When, for example, sound A and sound B of distinctly different loudness reach our ear at the same time, our ear possibly perceives only the stronger one (sound A in Fig. 11.6b). The deflection of the basilar membrane for the louder sound covers that for the smaller one, so that only the signal for the louder sound A is transmitted further.

11.2.3.2   Additional Tones

Different from the case of masked tones where air vibrations are present but no corresponding electric signals afterwards, there exists the opposite case where electric signals are created without vibration signals from the outside.

When sound consists of two strong tones A and A' of high frequencies f_A and $f_{A'}$ (combination tone), we hear not only the tones with pitches f_A and $f_{A'}$, but also the difference tone C (in Fig. 11.6c) having the pitch $f_{A'} - f_A$ (the first order difference tone), sometimes also the second order (cubic) difference tone $(2f_A - f_{A'})$ for $f_{A'} > f_A$. These difference tones can not be detected in the environmental air where the sounds A and A' are emitted, but exist only in our ear. This phenomenon was already reported by the organist G. Sorge in 1740, and later by the violinist G. Tartini. The difference tones are generated by nonlinear effects inside the cochlea. If both soprano and alto sing together in a duet with strong intensity, you may hear unexpected difference tones.

Another additional tone is the "missing fundamental": given a series of pure tones whose frequencies correspond to a harmonic series, one will hear an additional pitch near or at the fundamental frequency, even if there is no such tone at that frequency.

11.2.3.3   Cocktail Party Effect

Suppose that there are many people talking in pairs or in groups at a cocktail party. If you are standing in the hall alone, you may hear many people talking, but usually you do not know what they are talking about. When you have a conversation with somebody in such an environment, you can hear the voice of the person with whom you are talking clearly enough to understand the meaning. You hear many other voices at the same time, but these are a background noise for you. Such an effect is called the "cocktail party effect".

Not only at a cocktail party but also in a concert one can listen to a musical part (for example, the flute) separately from the total orchestra, if one wishes to.

There are many different tones coming to the ear and most of these vibration signals are converted to electric signals. They are still processed, just with less priority. Only the relevant tones are mapped in the primary auditory cortex. This phenomenon is caused by an effect of selective attention in the brain. Such auditory attention primarily occurs in the left hemisphere of the superior temporal gyrus (where the primary auditory cortex is located) [12]. But we do not know how. This is also a black box for us.

11.3 Absolute Pitch and Relative Pitch

11.3.1 What Are Absolute Pitch and Relative Pitch?

One other fascinating property of human speech perception is the ability to perceive absolute and/or relative pitch. Absolute pitch[7] is the ability of a person to identify any given tone frequency upon hearing without a reference tone. Such a person can also recreate a tone of the exact pitch so that he (or she) can sing a song exactly in the key indicated by the composer, or can reproduce a tone heard on a musical instrument immediately.

It is estimated that less than 0.1% of people possess absolute pitch [13]. In East Asia, where tone languages[8] (e.g. Mandarin, Thai, Vietnamese...) are spoken, however, more people have absolute pitch than in other areas where non-tone languages (e.g. English, French, Japanese...) prevail. Deutsch et al. reported that there are two different critical ages for obtaining absolute pitch. For tone language speakers it is the same age as the age of acquisition of their native language, while for non-tone language speakers it is the age for learning the second language [14]. It seems in many cases that the ability of absolute pitch is included in the mother tongue for the tone language speakers, while to train absolute pitch requires a big effort (like learning the second language) for non-tone language speakers.

---

[7]Sometimes the terms "perfect pitch" or "perfect absolute pitch" are used. They are differentiated from the "partial absolute pitch", which is interpreted as a short term memory of absolute pitch or as a limited range of the pitches recognized.

[8]A tone language, or tonal language, is a language in which words can differ in tones (like pitches in music) in addition to consonant and vowel quality.

How about musicians? Is it important to possess absolute pitch? Mozart is known to have demonstrated the ability to determine pitch at age 7 [15]. Also many famous musicians have such ability, for example, the violinist Itzhak Perlman or the cellist Yo-Yo Ma. Is it advantageous to have absolute pitch for playing musical instruments? If so, is it possible to train for obtaining absolute pitch?

Another variant of pitch perception ability is relative pitch. This signifies the ability of a person to identify a given tone by comparing it to a reference tone. Persons who possess relative pitch know the interval between a given tone and the reference. If they sing a song even without a reference tone, the melody is correct: They use the first tone of arbitrary pitch as a reference tone for the next one.

11.3.2   Solfège

Before we discuss the importance (or non-importance) of absolute pitch for musicians, we introduce solfège. Solfège is a method of naming pitches by using syllables. Actually the names are taken from the Italian expressions for natural notes (see Chap. 2). Instead of naming a C major scale as C-D-E-F-G-A-B-C, it is named as do-re-mi-fa-so (or sol)-la-si (or ti)-do. Solfège is often used by singers, because it is easier to sing the syllables than the letters.

There are two different systems in solfège: the movable 'do' system and the fixed 'do' system.

In the fixed 'do' system 'do' is always C, 're' always D, etc., regardless of the key. This is actually the same system as introduced in Table 2.1 in Chap. 2. Each natural note has an individual name, which corresponds to its pitch (frequency).

In the movable 'do' system, on the other hand, the 'do' is always the tonic, regardless of the scale. Therefore, 'do' is D in D major, 'do' is E in E major, and so on. For this system not the pitch but the interval from the fundamental tone is essential.

Let us sing the German children song "Hänschen klein" with solfège. The beginning of the note in C major (top) and D major (bottom) are shown below. The names of each tone are written in the movable 'do' system.

Hänschen klein in C major (top) and in D major (bottom)

For the people who have no absolute pitch, but relative pitch (like KT), it does not matter so much whether this song is sung in C major (G E E F D D...) or in D major (A F♯ F♯ G E E ...). And it is not a problem to sing "so mi mi fa re re..." in both cases.

One day, when KT sang "so mi mi fa re re ..." without any reference pitch during preparing lunch, one woman shouted "Stop singing! Otherwise I would get crazy." "Hee? What's the problem?" "You sang completely wrong. Your 'so' is not 'so'." (She meant that KT's 'so' is not 'G'.) During the lunch conversation KT got to know that this woman possesses absolute pitch and that "do re mi fa so la si do" works for her only in the fixed 'do' system. She cannot transpose even a simple melody like "Hänschen klein" spontaneously from C major to D major. On the contrary, KT can sing "Hänschen klein" in any key without any conflict with the written sheet. Just she does not know in which key she is singing.

11.3.3 How Do Absolute Pitch and Relative Pitch Work?

It seems that the people who possess absolute pitch have a "look (hear)-up" table of acoustic frequencies and the corresponding name in their brain (like Table 2.2 in Chap. 2). If they hear a sound of a certain frequency, they find its name from this list immediately. If they name a certain tone, then they can sing the corresponding pitch which they find in the table.

For relative pitch it is more complicated. When they start singing "Hänschen klein" with the tone G by chance, they sing E as the next tone without knowing it is E. When they start with the tone A by chance, they sing F♯ as the next tone without knowing it is F♯. They only know that "mi" should come after "so" (in the movable 'do' system). It means that they know that the second note is 3 half tones lower than the first note. How does this work even without knowing solfège?

People who possess relative pitch do not have a table of acoustic frequencies and the corresponding name, but they must have a small, but very fast calculator and a universal table (much smaller than the table for absolute pitch) in their brain. It may work as follows: when a relative-pitch person hears one tone, and an absolute-pitch person tells him that this tone is F, the relative-pitch person's brain immediately constructs the C major scale, so that this person knows not only the interval between C and F but also any other interval. Their brain possesses a very simple movable (or universal) table shown in Table 11.1. This table can be used regardless of the frequency of 'do'. Therefore, once the relative-pitch person knows one reference tone (in this case F), this person can identify other tones heard by calculating the frequency ratio.

Table 11.1 Small movable table for relative pitch

Next tone	do	re	mi	fa	so	la	si	do
Ratio	1/1	9/8	5/4	4/3	3/2	5/3	15/8	2/1

Let us denote the frequency of the heard tone X as f_X.

If the person obtains information that X is the note F, then C is 5 tone intervals (perfect fifth) lower. Therefore, the frequency of C can be calculated with the frequency ratio according to Table 11.1:

$$f_C = \frac{3}{4} f_F.$$

(Note that this argument holds for the European musical environment. If the person has grown up in a pentatonic environment, then the most familiar pentatonic scale for this person should be constructed as a movable table.)

Let us come back to the German children song "Hänschen klein". Some relative-pitch people start singing with C by chance, others with D (one tone higher) or with Bb (one tone lower). They know the interval between the first and the second tones is the minor third, the second tone and the third tone are unison, …, and so on, regardless of the starting pitch. The people who start singing with G (without knowing it is G) immediately construct a C major scale (without knowing it is the key in C). Once the corresponding scale is made, they know the pitch for any interval. They calculate next pitches:

$$f_{C2} = \frac{3}{2} \cdot \frac{4}{5} f_{C1}, \quad f_{C3} = f_{C2}, \quad f_{C4} = \frac{3}{2} \cdot \frac{3}{4} f_{C1}, \quad \ldots,$$

where f_{CN} is the frequency of the tone at the Nth position in the C major key. (N corresponds to the numbers in red above the note in "Hänschen klein".)

In this case, if they start singing with D, the D major scale is constructed. The following calculation proceeds in exactly the same way.:

$$f_{D2} = \frac{3}{2} \cdot \frac{4}{5} f_{D1}, \quad f_{D3} = f_{D2}, \quad f_{D4} = \frac{3}{2} \cdot \frac{3}{4} f_{D1}, \quad \ldots \quad \ldots\ldots$$

Since the starting pitches (f_{C1} for the first case and f_{D1} for the second case) are different, the f_{CN} and f_{DN} are also correspondingly different. So, one can sing "Hänschen klein" with the correct melody in various keys.

This is our actual thinking how the music works in the brain of relative-pitch people, which is for us currently the most plausible explanation. Some neuroimaging research supports our thinking: it reveals that absolute pitch people use long term memory, while relative pitch people activate working memory, when listening to transposed tone sequences [16].

11.3.4 Piano Versus Violin, Bells Versus Monochord

The movable 'do' system is very practical for those people who possess only relative pitch. If you know one melody, you can transpose it to different keys without changing the name of each tone. Or, on the other hand, when KT plays the Piano sonata in C major of Mozart, she sings the theme in her head "do- mi so si- do re do..." as shown in the notes on the top of the following graphics. Later the same melody appears in F major (notes on the bottom), and she also sings "do- mi so si- do re do...". KT wonders, how somebody who has absolute pitch would sing that: maybe "fa- la do mi- fa so fa..."?

Mozart Sonata in C major KV 545, top: 1st–4th bars, bottom: 42nd–45th bars

This sonata is a piano piece. The upper notes are played only with white keys, while one black key has to be used for the lower notes. Therefore, the fingering should be changed, and this is one of the difficulties for beginners. But if this part is played with a violin, the fingering for the upper notes in C major and the lower notes in F major is, in principle, identical (when played on a violin string with the appropriate position of the left hand), because the frequency ratio of each note to the first note (this corresponds to the interval) is identical. In this sense the piano is an instrument of absolute pitch, while the violin is an instrument of relative pitch.

One further consideration: Both Greece and China developed similar tonal scales based on frequency ratios of 2:3, 3:4, 4:5, etc (see Chaps. 1, 2 and 5). Pythagoras did not define the length of his "monochord", therefore, his scales could change the root. The Chinese constructed many bells[9] to conserve correct pitches as shown in Fig. 11.7 [17]. Relative pitch was essential for Pythagoras, while for the Chinese absolute pitch was important. Probably this difference is associated with their languages: Greek is a non-tone language and Chinese is a tone language.

Fig. 11.7 Chinese bells, Bianzhong of Marquis: traditional Chinese bells found in 1978 in the tomb of Marquis Yi of Zeng. The Bianzhong bells were made in 433 BC (CC BY-SA 3.0 by Zzjgbc)

---

[9]There are 64 Bianzhong bells in total. Each bell can play two tones with three half-tone intervals. The tonal range of the bells covers the interval from C2 to D7. In the middle area of the tonal range, it can play all twelve half tones.

Toward the end of this section we like to emphasize that there are advantages and disadvantages of absolute pitch and relative pitch for learning, playing and enjoying music. If you have absolute pitch, take its advantages. If you have relative pitch, take its advantages. To have absolute pitch is not essential for music. We are sure that it is much better for children to enjoy nice music than to train absolute pitch.

11.4 Auditory Illusion

Beyond considering aspects of pitch perception we now proceed to another type of perceiving sound input—the auditory illusions. These deal with false perception of a real sound input—just like also known from visual illusions. As examples we explain binaural beats and octave/scale illusions. Furthermore, we introduce pitch circularity, which is equivalent to the Penrose or Escher's staircase. Anomalies such as tinnitus or musical hallucinations are also considered.

And we are asking the question, whether we are facing problems of physics, physiology, brain dynamics or music—or all of these disciplines.

11.4.1 Binaural Beats

When two sounds of slightly different frequencies are played simultaneously, one hears a tone of the average frequency of the original two sounds, the amplitude of which oscillates with the difference of frequencies of these two sounds. Such phenomena are caused by interference of two waves and are called "beats" (refer Sect. 3.1.3).

Different from normal beats, a binaural beat is an auditory illusion. When two sinusoidal tones (frequencies; lower than 1500 Hz, difference between two tones: less than 40 Hz) are presented to a listener dichotically (one tone to the right ear and the other to the left ear), the listener hears a beat, as if these two waves coexist in the air.

For example, if a tone of 440 Hz is put into the right ear, and a tone of 450 Hz in the left ear, the listener perceives the auditory illusion of a third tone of frequency 10 Hz, in addition to the two tones presented to each ear. This third sound is called a binaural beat.

The binaural beat is caused in certain parts of the brain, where auditory signals from each ear are integrated [18].

11.4.2 Octave Illusion and Scale Illusion

These illusions were found by Diana Deutsch [19, 20].

When both the left and right ears receive separately (through stereo headphones) two octave tones alternating between high and low, most of the people perceived the high tone in the right ear and the low tone in the left ear and vice versa as shown in Fig. 11.8 left.

There was a statistically significant strong tendency for the right-handed subjects to hear the high tone in the right ear and the low tone in the left. For the left-handers there was no clear preference. Deutsch considers this difference between right and left handers to be related to the hemispheric dominance of handedness.

Later it was found that the illusion is not limited to the octave, but also scales in the range of a minor third to an eleventh [21].

Similar experiments were carried out with the repetitive presentation of the C major scale with successive tones alternating from ear to ear: When a component of the ascending scale was heard by one ear, a component of the descending scale was perceived in the other, and vice versa (see Fig. 11.8, right). Similar to the case of the octave illusion, illusory perception correlates with the handedness of the listener. Different from the octave illusion, tones are continuously heard by both ears. No tones are ignored. So far there is no explanation for this difference.

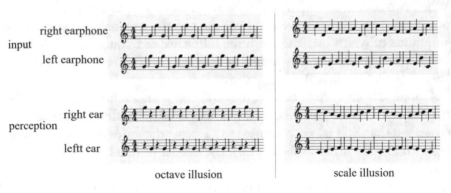

Fig. 11.8 Typical acoustic illusions, left: octave illusion, right: scale illusion

An interesting example of the scale illusion (or maybe not) is the Symphony No. 6 in B minor Op. 74 (Symphony Pathétique) of Pyotr Ilyich Tchaikovsky. One hears at the beginning of the final movement of this symphony the melody

Fig. 11.9 Top: 1st and 2nd violin part of the beginning 2 bars of the final movement of the Pathétique, bottom: perception [22]

shown in the third line from the top of Fig. 11.9 . However, the musical notes actually played are shown in the upper two lines (the 1st violin plays the first line, the 2nd violin the second line). If the 1st violinist practices this part alone, he may not recognize the famous melody. The same holds for the 2nd violinist. Only when they play together, and if they are good enough to listen to the other part, they can notice the melody, which is a sequence of the higher tones.

Let us suppose that the orchestra seating arrangement is of the old German style: The 1st violins sit on the left side of the front stage, and the 2nd violins on the right side of the stage (see Chap. 10). When the music starts, one hears the melody. According to Deutsch's theory on the scale illusion, the audience (mainly righthanded) should hear the melody from the right side where the 2nd violins are playing. However, from her "Pathétique experiment" the audience perceives as if the 1st violins play the melody. This could be probably because we are biased from the fact that in most of the cases the 1st violins play the melody. We are sure that Tchaikovsky did not think about the scale illusion, but knew the effect of the spatial separation of the two violin parts, which makes the sound richer and bigger, as the composer Michael A. Levine suggested.

Coffee Break

If many composers would have written scores in which a theme and accompaniment warft back and forth between the 1st violin and 2nd violin, as is seen in the last movement of the Pathétique, the German seating arrangement would have been used more often.

11.4.3 Pitch Circularity and Tritone Paradox

11.4.3.1 Pitch Circularity

Roger Shepard demonstrated how pitches rise (or fall) endlessly [23]. He introduced a pitch class, meaning a set of pitches in octave intervals. For example, C1, C2, C3, …Cn belong to the pitch class C, while D1, D2, …Dn to the pitch class D. Shepard constructed 12 pitch classes artificially, each of which consists of a set of pure tone components separated by octaves, and has a bell-shaped spectral envelope (when the abscissa is $log_{10}f$, where f is frequency), as shown in Fig. 11.10 for the case of the pitch class C.

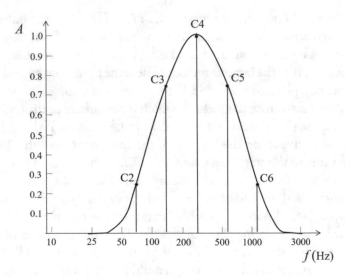

Fig. 11.10 Spectral structure of Shepard tones C

For the question which is higher, C4 or D3, the right answer "C4" comes immediately. On the other hand, answering the question which is higher, the pitch class C or the pitch class G, is not so clear because of the nature of the pitch class: each pitch class consists of several tones separated by octaves with different amplitudes. When listening to the sound of pitch class C, one is not sure whether the sound heard is C3 or C4 or another C; listening to the sound of pitch class G, one is not sure, whether the sound heard is G3 or G4 or another G.

Shepard asked listeners who heard two pitch classes played in succession, whether they were ascending or descending. He found that this depends on the distance in the pitch class circle shown in Fig. 11.11 (left). For example, if the pitch class C and then the pitch class D are played sequentially, listeners heard this sequence always as ascending, but the sequence of the pitch class C and A♯ as descending. The shorter distance on the circle from C to D is clockwise, while that from C to A♯ is couterclockwise.

This is completely different from listening to any sound on the piano, violin or other musical instruments. In this case we hear tones ascending when they are played clockwise in the 12 half-tone circle (Fig. 11.11, right), and descending when played counterclockwise, and this is independent of the distance in the circle.

From these results of perception Shepard got the brilliant idea to produce an endlessly rising pitch by playing pitch classes clockwise, and an endlessly falling pitch by playing pitch classes counterclockwise. These endlessly rising or falling pitches remind us of the Penrose staircase [24] shown at the center of the pitch class circle Fig. 11.11, as well as Escher's graphics.

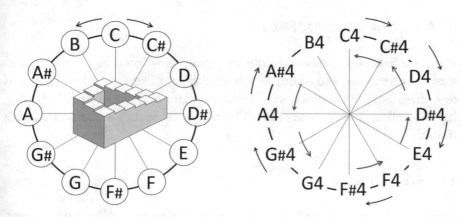

Fig. 11.11 Left: pitch class circle consisting of the 12 pitch classes within the octave (at the center of this circle the Penrose staircase is shown; public domain by Sakurambo); right: half-tone sequence from C4 to B4 on a circle (Note that these circles are different from the circle of fifth described in Chap. 4)

11.4.3.2 Tritone Paradox

What will happen, if a pitch class sound and its tritone are played in succession? This was also a question of Deutsch et al. [25].

The tritone is a chord encompassing three adjacent whole tones. Its intervals are the diminished fifth or augmented fourth. Since it is an extremely dissonant interval, one calls it the devil's interval (see Chap. 3). In the equal temperament (see Chap. 4) the tritone cuts the octave into two halves (of 600 Cent each). Therefore, the interval between the higher tritone (half an octave higher) and the lower tritone (half an octave lower) from any root tone is always an octave. In Fig. 11.11 (left) each tritone pair locates at opposite positions on the circle.

Results: some test people say it is descending, and others say ascending to the same tritone pairs of the pitch class. Deutsch et al. found a strong connection between listeners' perception and certain differences in linguistics.

When KT heard Deutsch's lecture on a WEB site (at that time she knew neither about the tritone paradox nor pitch class tones), she had difficulties to answer whether sound of two successions is ascending or descending. Actually she heard funny mixtures of higher tones and lower tones. But there was no option to answer "ascending and descending" and no chance to comment about it. We wonder whether Deutsch's results could be different, if she offers one more option "both ascending and descending".

11.4.4 Not Illusions, But...

There are some other anormal auditory perceptions (tinnitus, earworm, musical hallucinations, and so on) which are different from auditory illusions as described above. The borderline of such anormalities is not so clear. Nevertheless, we will introduce some of them briefly, as far as they are associated with music.

11.4.4.1 Tinnitus

Subjective tinnitus[10] is the hearing of sound when no external sound is present. The sound heard in ears is often described as a ringing, clicking, hiss, or roaring with high pitch or low pitch.

Tinnitus is a symptom and not a specific disease, and 10–15% of people suffer from it. The causes of tinnitus are exposure to loud noises, aging, taking ototoxic drugs, Ménière disease,[11] middle ear infection and others.

---

[10] Subjective tinnitus is the perception of sound in the absence of an acoustic stimulus and is heard only by the person himself. Most of perceived tinnitus is subjective. Objective tinnitus is uncommon and results from noise generated by structures near the ear. Sometimes such tinnitus is loud enough to be heard by the examiner.

[11] Ménière disease is a disorder of the inner ear that is characterized by episodes of feeling like the world is spinning, ringing in the ears (tinnitus), hearing loss, and a fullness in the ear.

The Bohemian composer Bedřich Smetana (1824–1884) got totally deaf when he was 50 years old and since then suffered from tinnitus. When he composed his String Quartet No. 1 in E minor "From my Life" in 1876, he inserted the tinnitus sound, which he heard, into the 4th (last) movement. It is a tone with high pitch (E7 ≈ 2794 Hz) followed by an uneasy passage.

This work presents Smetana's own life. Especially the 4th movement is a "program-music". He used an "almost" physical material to express his sadness for losing his hearing ability. One more notable feature of this quartet includes a prominent viola solo at the very beginning of the first movement [26]).

11.4.4.2 Earworm

An earworm is a song or a melody that keeps repeating in one's mind. It is also called a brainworm, sticky music, stuck song syndrome, or IMI (Involuntary Musical Imagery). It is generated without external stimuli.

Quite many people have an earworm. For example, according to a survey of Finnish internet users, more than 90% of them hear involuntary earworms at least once in a week [27].

Catchy music or commercial songs which are played repeatedly in television or supermarkets appear often involuntarily as an earworm [28]. When such earworms visit one's head often and/or stay long, this person is irritated and disturbed. On the contrary, KT has a nice earworm as shown below.

KT's earworm: the 3rd movement of the Piano Trio No. 39 in G major of
Joseph Haydn. This movement is called "Rondo, in the Gypsies style"

This passage consists of repeating cycles with a regular but very fast (presto), vivid rhythm, and the passage itself appears many times. She always enjoys playing this piece, and the earworm is also always welcome. However, it does not last for a long time.

Sometimes the word "earworm" is used in the broad sense, for the case when melody or songs are imagined by one's will. Musicians use this worm voluntarily to rehearse or aid reproduction of music.

Intermezzo

The word "earworm" is a word directly translated from the German word "Ohrwurm", which has had this meaning since the mid-20th century. The earliest known English usage is found in Desmond Bagley's 1978 novel Flyaway, where the author points out this translation as his invention.

> I fell into a blind, mindless rhythm and a chant was created in my mind what the Germans call an 'earworm' something that goes round and round in your head and you can't get rid of it. One bloody foot before the next bloody foot.

11.4.4.3 Musical Hallucination

A musical hallucination is the perception of music, when none is playing. It belongs to the auditory hallucinations, which are defined as the conscious experience of sounds that occur in the absence of any actual sensory input. It is reported that only 0.16% of the people (in a cohort study of 3,678 individuals) suffer from this disorder [29]. However, the percentage could be much higher: Many people might not talk about such hallucinations, because they are afraid to be treated as a psychotic.

O. Sacks reports many cases [30]. Some sound loud and hard, others not so loud and soft. In many cases patients hear fragments of music, which they heard in childhood. Some people can manage so that they hear a more consistent melody, or even that they can shift from one music to another music.

In many cases musical hallucination is caused during losing one's hearing ability. It looks like a person, who becomes unemployed, pretending to go to work everyday as before: One day he goes to a film theater, next day to a zoo, and then to a shopping center, and further without any sense. Brain cells which are responsible for auditory activities are trying to keep active, creating own signals, although they do not get signals from outside.

Instead of asking "Why do hallucinations occur?" the Polish neurophysiologist J. Konorski asked "Why do hallucinations not occur all the time?" [31]. His idea:

> The mechanism producing hallucinations is built into our brains, but it can be thrown into operation only in some exceptional conditions.

is supported by good evidence from PET studies [32].

Our discussion of several auditory illusions underlines convincingly, how efficient an interdisciplinary approach can be for bridging the gap between physics and musical sensation, when considerations from physiology and brain dynamics are included.

References

1. L. Chittka, A. Brockmann, Perception space - The final frontier. PLoS Biology **4**(3), e137 (2005)
2. Bone conduction: How it works, http://www.goldendance.co.jp/English/boneconduct/01.html
3. H. Helmholtz, *Théorie Physiologique de la Musique* (Victor Masson et Fils, Paris, 1868)
4. G. von Békésy, *Experiments in Hearing* (McGraw-Hill Inc., New York, 1960)
5. J.S. Oghlai, The cochlear amplifier: augmentation of the traveling wave within the inner ear. Curr. Opin. Otolaryngo. **12**, 431–438 (2004)
6. S.S. Narayan, A.N. Temchin, A. Recio, M.A. Ruggero, Frequency tuning of basilar membrane and auditory nerve fibers in the same cochleae. Science **282**, 1882–1884 (1998)
7. J. Zheng, W. Shen, D.Z. He, K.B. Long, L.D. Madison, P. Dallos, Prestin is the motor protein of cochlear outer hair cells. Nature **405**, 149–55 (2000)
8. U. Müller, Cadherins and mechanotransduction by hair cells. Curr. Opin. Cell Biol. **20**, 557–566 (2008)
9. P. van Dijk, D.R.M. Langers, Mapping tonotopy in human auditory cortex, in *Basic Aspects of Hearing*, ed. by B.C.J. Moore, R.D. Patterson, I.M. Winter, R.P. Carlyon, H.E. Gockel (Springer, Cham, 2013), pp. 419–425
10. A. Ogata, Science of do, re, mi. Lecture material in TSS Culture University held on 13.03.2012; https://home.hiroshima-u.ac.jp/masters/TSS-gakumon-sanpo/23-11-ogata.pdf
11. E. Meyer, E.-G. Neumann, *Physikalische und Technische Akustik* (Friedrich Vieweg und Sohn, Braunschweig, 1974), pp. 221–223
12. N. Wood, N. Cowan, The cocktail party phenomenon revisited: How frequent are attention shifts to one's name in an irrelevant auditory channel? J. Exp. Psychol. Learn. **21**, 255–260 (1995)
13. I.J. Profita, T.G. Bidder, Perfect pitch. Am. J. Med. Genet. **29**, 763–771 (1988)
14. D. Deutsch, T. Henthorn, E.W. Marvin, H.-S. Xu, Absolute pitch among American and Chinese conservatory students: Prevalence differences, and evidence for a speech-related critical period (L). J. Acoust. Soc. Am. **119**, 719–722 (2006)
15. D. Deutsch, The enigma of absolute pitch. Acoust. Today **2**, 11–18 (2006)
16. D.J. Levitin, S.E. Rogers, Absolute pitch: Perception, coding, and controversies. Trends Cogn. Sci. **9**, 26–33 (2005)
17. S. Shen, Acoustics of ancient Chinese bells. Sci. Am. **256**, 94 (1987)
18. G. Oster, Auditory beats in the brain. Sci. Am. **229**, 94–102 (1973)
19. D. Deutsch, An auditory illusion. Nature **251**, 307–309 (1974)
20. D. Deutsch, Two-channel listening to musical scales. J. Acoust. Soc. Am. **57**, 1156–1160 (1975)
21. A. Brancucci, C. Padulo, L. Tommasi, "Octave illusion" or "Deutsch's illusion"? Psychol. Res. Psychol. Fo. **73**, 303–307 (2008)

22. D. Deutsch, Grouping mechanism in music, in *The Psychology of Music* (Elsevier, San Diego, 2013), pp. 183–248
23. R.N. Shepard, Circularity in judgements of relative pitch. J. Acoust. Soc. Am. **36**, 2346–2353 (1964)
24. L.S. Penrose, R. Penrose, Impossible objects: a special type of illusion. Brit. J. Psychol. **49**, 31–33 (1958)
25. D. Deutsch, T. North, L. Ray, The tritone paradox: correlate with the listener's vocal range for speech. Music Percept. **7**, 371–384 (1990)
26. E. Heimeran, B. Aulich, *Das stillvergnügte Streichquartett*, 13th edn. (Ernst Heimeran Verlag, München, 1956), pp. 103–104
27. L.A. Liikkanen, Music in everymind: Commonality of involuntary musical imagery. in Proceedings of the 10th international conference on music perception and cognition (ICMPC10) Sapporo, Japan (2008) ed. by K. Miyazaki, Y. Hiraga, M. Adachi, Y. Nakajima and M. Tsuzaki, pp. 408-412
28. K. Jakubowski, S. Finkel, L. Stewart, D. Müllensiefen, Dissecting an earworm: melodic features and song popularity predict involuntary musical imagery. Psychol. Aesthet. Crea. **11**, 122–135 (2017)
29. S. Evers, T. Ellger, The clinical spectrum of musical hallucinations. J. Neurol. Sci. **227**, 55–65 (2004)
30. O. Sacks, *Musicophilia: Tales of Music and the Brain* (Picador, London, 2012)
31. J. Konorski, *Integrative Activity of the Brain: An Interdisciplinary Approach* (University of Chicago Press, Chicago, 1967)
32. T.G. Griffiths, Musical hallucinations in acquired deafness: Phenomeology and substrate. Brain **123**, 2076–2076 (2000)

12

Room Acoustics

If a string ensemble chooses a meadow on which to play chamber music, this may not be such a good idea, for sound may trail away into the middle of nowhere. A brass ensemble will do better, because it has a more "frontal" radiation pattern and thus remains well audible. Suppose now that the strings move into a large church hall, the results of their effort may suffer from the fact that the church walls reflect sound as reverberation or echo, which is intense and decays slowly. Consequently, the faster musical passages get blurred. Thus: an adequate room is required for a musical performance (better more than a concert shell for a promenade concert).

Nevertheless, in antique times performances under open sky were very common. Amphitheaters still used as witnesses for early cultural events as shown in Fig. 12.1, or even—if restored—for modern performances, were beautiful structures in attractive settings, but had the disadvantage that there were reflections occurring at the regularly arranged seating rows. These prove to have a detrimental effect on the acoustics.

As a remark: The top of pyramids built during the dominance of Maya civilization could often be reached by climbing up long staircases (Fig. 12.2). As an echo to clapping one's hands at the basis of the pyramid one can hear a sound which resembles the call of a local bird. Similarly, a strong echo can be produced when standing in front of a large and even side of a pyramid: Sound is reflected by several hundred meters and the hand clapping is perceived as a gun shot [1]. Such an echo is particularly remarkable when standing inside an area for Maya ball games.

For many centuries up to today closed rooms are being preferred for musical performance. These have a significant influence upon the acoustics of emitted sound.

© Springer Nature Switzerland AG 2021
K. Tsuji et al., *Physics and Music*,
https://doi.org/10.1007/978-3-030-68676-5_12

Fig. 12.1 The antique theater of Aptera situated on a rock plateau near Chania, Crete. Built in hellenistic times and partially reconstructed (CC BY-SA 4.0 by Wikifreund)

Fig. 12.2 Left: Tikal Temple I, seen from the northwest as steep-sided (CC BY-SA 2.5 by Aquaimages), right: stepped pyramid with a central staircase that rises from a flat, grassy area to a temple doorway at the top (CC BY-SA 2.5 by Raymond Ostertag)

12.1 Concert Halls

To obtain an overview about concert halls is an unfeasible endeavor in the framework of this chapter. There is a plethora of halls for performing and presenting music, greatly differing in size, architecture, style and equipment, décor, and other "internal parameters" motivated by acoustical aspects and often hidden from the spectator.

We show in Figs. 12.3 and 12.4 two examples exhibiting significantly contrasting internal designs.

The arrangements for orchestra and audience, room shape, wall outfit, and other properties are very different. Which hall would now be better fitted for an orchestra? What kind of criteria should be applied?

Main words should come from architects and from scientists working in room acoustics (and they will often quarrel). An interdisciplinary topic for physics and arts!

Fig. 12.3 Auditorium of the Bolshoi Theatre in Moscow, in 2014 (CC BY 3.0 by Dmitri Dubinsky)

Fig. 12.4 Ceiling of Culture Palace (Tel Aviv) concert hall is covered with perforated metal panels (CC BY-SA 3.0 by Etan J. Tal)

12.2 Principles of Room Acoustics

This is a multi-parametric three-dimensional system which can be analyzed by observing different spatial and temporal scales. One can apply a theoretical approach based on the wave equation of sound propagation, if the wavelength of sound is of the order of the room dimension, and if room resonances play a significant role. Or, if the wavelength is very short in comparison to the room dimensions, the geometrical approach of sound beams is appropriate, which propagate along straight lines and are reflected by the walls. Furthermore, if the number of superposing direct and reflected beams becomes very large, only a statistical approach about the sound field becomes effective.

Intermezzo

If you sing a melody in the bathtub or under the shower, this usually happens in a small room. With the walls being covered with tiles, about 98% of the energy of any acoustical signal in your bathroom will be reflected. The three-dimensional room may have geometrical dimensions 90 cm × 90 cm (at the bottom) and 240 cm in height. In the vertical direction we consider this room as a ducked flute or an organ pipe with fundamental frequency twice its length. Division by the speed of sound in air of 340 m/s leads to a frequency of 70 Hz, together with harmonics with frequencies 140 Hz, 210 Hz, 280 Hz, 350 Hz etc.

Singing "O sole mio…" may start with a frequency around 350 Hz, which is a resonance frequency for the vertical direction. However, also transversal waves are possible, so that a threefold infinite sequence of resonances occurs. These are, in principle, determined by geometry, sound speed and natural numbers in groups of three. Their strength is largely influenced by the reflectivity of the walls.

The result of singing will thus not be the enjoyment of hearing one's own voice, but the strongest modes in the resonator where the bathtub is located.

As an example: wavelength λ of a tone of 1000 Hz is about 34 cm. For applying geometrical acoustics the reflecting surfaces should be larger by 2–3 times, i.e. 1 m².

12.3 Wave Theory

For describing the sound field in a cubic room we start from the wave equation for loss-free propagation, conveniently formulated for the velocity potential $\Phi(x, y, z)$

$$\Delta\Phi = \frac{1}{c^2}\frac{\partial^2\Phi}{\partial t^2} \tag{12.1}$$

From Φ one derives sound pressure as

$$p = \rho \frac{\partial \Phi}{\partial t} \tag{12.2}$$

and sound particle velocity as

$$\mathbf{v} = -grad\Phi. \tag{12.3}$$

The velocity potential must be a solution to the wave equation and also obey boundary conditions. As a good approximation to real conditions we assume reflecting walls, i.e. the velocity component normal to the wall disappears.

If this room is excited by a frequency of $f = \omega/2\pi$, the velocity potential will also have the same frequency in each point:

$$\Phi(x, y, z) = \Phi_\omega(x, y, z)e^{i\omega t}. \tag{12.4}$$

Inserting this expression into the wave equation leads to:

$$\Delta\Phi_\omega = -\frac{\omega^2}{c^2}\Phi_\omega = -k^2\Phi_\omega. \tag{12.5}$$

This partial differential equation together with the required boundary conditions can be only solved for simply formed volumes, where the variables can be separated.

The solution for a cube-shaped room with side lengths l_x, l_y and l_z ($l_x > l_y > l_z$) would then be:

$$\Phi = \hat{\Phi}\left(\cos\frac{n_x\pi x}{l_x}\cos\frac{n_y\pi y}{l_y}\cos\frac{n_z\pi z}{l_z}\right)e^{i\omega t}. \tag{12.6}$$

The numbers n_x, n_y and n_z are natural numbers with values larger than or equal to zero. This ansatz for Φ solves the wave equation (Eq. 12.1) for discrete values of the frequency f_r. These eigenfrequencies of the room are determined by the geometrical shape of the room.

Each triple of numbers signifies an eigenfrequency of the room according to

$$f_r^2 = \left(\frac{c}{2}\right)^2\left[\left(\frac{n_x}{l_x}\right)^2 + \left(\frac{n_y}{l_y}\right)^2 + \left(\frac{n_z}{l_z}\right)^2\right]. \tag{12.7}$$

These numerals count the number of planes of pressure nodes to be passed when traversing the room parallel to any of the coordinate axes. An overview of the quantity of eigenfrequencies can be obtained by inserting the coordinates $(\frac{n_x c}{2l_x}, \frac{n_y c}{2l_y}, \frac{n_z c}{2l_z})$ as grid points into a cartesian coordinate system with axes f_x, f_y and f_z.

Their frequency is just the distance of the grid points to the origin of this system. With this representation one can make an estimate of the number N of eigenfrequencies being smaller than a certain upper bound f_{gr}. Just divide the volume of the room by the volume of an elementary cube existing for each grid (frequency) point. Then include grid points on the room axes and the coordinate surfaces and get an expression for N such as shown in Fig. 12.5 [2].

In a large concert hall one may disregard, in a first step, the precise form of this hall and consider a cube with equal volume. If that volume is 75,000 m³, one finds in the small interval between 1000 and 1001 Hz about 25,000 eigenfrequencies. Thus, there are so many eigenfrequencies in any frequency interval, that amplification of any specific interval by resonance is not to be expected. All eigenfrequences can be considered as standing waves being reflected between the walls. There is a multitude of modes formed this way. These can be selectively damped by applying absorbing materials on the walls in question. But in principle all possible propagation directions of these waves have equal probability and statements on the sound field can only be made in a quite general way. Consequently, a statistical approach is called for.

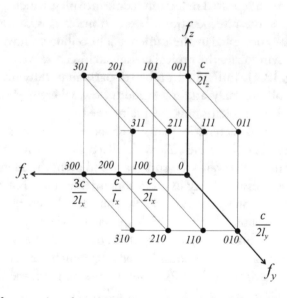

Fig. 12.5 Eigenfrequencies of a cubic volume. The length of the radius vector from the origin to a grid point measures the corresponding frequency, its direction the propagation in space

12.4 Statistical Room Acoustics

Instead of superposing all these many eigenfrequencies, it turns out to be much easier to apply statistical methods to characterize the resulting sound field. We seek for general laws to describe the onset and decay of such sound fields, independent of the particular shape of a room. Relevant are mainly its volume V, the total surface A of the walls and their absorption properties with coefficient α.

12.4.1 Reverberation

This term describes the persistence of sound after the sound is produced. It is created when a musical signal is reflected, causing numerous reflections to build up and then decay, as the sound is absorbed by the surfaces of objects in the space—including walls, furniture, people and air. It becomes noticeable when the sound source stops but the reflections continue, their amplitude decreasing until the signal "fades away".

For the reflections to be considered in this context we show in Fig. 12.6 how, after the first direct sound signal has arrived at the detector site, early, secondary waves reflected at the walls—smooth or rough—arrive next (for details see Sect. 12.5), and then repeated reflections come into play, which become weaker in time and cause the reverberation. The bottom in this cartoon is flat, but would be normally occupied by the audience. The ceiling is drawn in a sawtooth manner, which symbolizes the typical ceiling architecture which is complex in nature (see Fig. 12.4). Individual orchestra parts may thus merge with each other, melodic phrases may attain a smooth flow, whereby distinctiveness of short breaks and rhythmicity must be maintained.

A most efficient characteristic quantity of spacious rooms is the concept of the so-called "reverberation time", T_r, which is a major parameter for evaluating the quality of music halls. With the end of a sound event the energy density and thus the acoustic pressure decay in an exponential fashion. Following Wallace Clement Sabine [3] one does not merely consider the corresponding decay time but the time interval, after which the energy density is reduced to its 1-millionth value. This interval corresponds roughly to the time, after which our hearing would perceive any sound, even after which the real sound is being switched off. For a drop of the SPL (sound pressure level, see Sect. 6.2.3) by 60 dB one calculates the reverberation time T_r as

$$T_r = 0.163 \frac{V}{A} \frac{1}{\alpha} \tag{12.8}$$

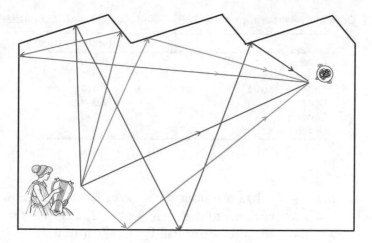

Fig. 12.6 Various reflections, red: direct, green: 1 reflection, blue: 2 repeated reflections, purple: 3 repeated reflections

where 0.163 was obtained empirically by Sabine and has the dimension of sm^{-1}, V is the volume of the room, A is the total area of the concerning surface, and α is the absorption coefficient. This coefficient describes the portion of absorbed/impinging sound energy and depends on the angle of incidence.

An idealized reverberation curve and an example (reverberation curves for a musical piece) are shown in Fig. 12.7.

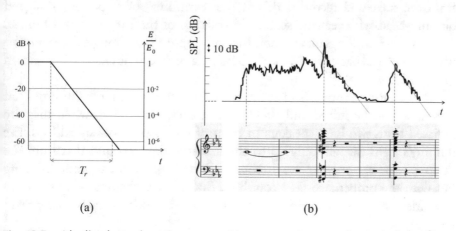

(a)   (b)

Fig. 12.7 a Idealized reverberation curve with E = sound energy density in [Ws/m³], E_0 value for $t = 0$, **b** reverberation curves for a musical piece with 2 accentuated chords and a rest in between

Table 12.1 Optimum reverberation times T_r (s) (500–1000 Hz) for various musical genres

	Optimum T_r (s)
Cabaret	0.8
Lecture, play	1.0
Chamber music	1.4
Opera	1.3–1.6
Concert	1.7–2.0
Organ music	2.5

Thus, the time T_r of a hall is proportional to the linear room dimensions V/A and inversely proportional to the coefficient α. T_r depends strongly on the properties of the walls, the bottom and the ceiling of a hall, which are frequency dependent and influence the timbre of a sound. Another important factor is the audience, i.e. the quality of the seats and the number of listeners.

If T_r is very long, then speech perception will be quite bad; for music so-called "dry" rooms are not preferable, but a certain interval of T_r is desired. Pleasant and distinct sound requires reflections to be sufficiently diffused (Table 12.1).

In the mid 1970s a noteworthy study of concert halls was performed by correlating subjective data with objective (geometric and acoustic) parameters of the halls. Schroeder and his team [4] visited 22 halls scattered throughout Europe, where they produced tape recordings of a symphony orchestra played from the stage and recorded this at the "eardrums" of a specially designed dummy-head with realistic surface impedance of the ear canal and pinnae. These stereo recordings of, why not, Mozart's Jupiter Symphony, 4th movement were later played back over two loudspeakers in an anechoic chamber[1] as shown in Fig. 12.8.

For recreating a sound field at the two ears of the listener, the dummy-head signals should be mixed and prefiltered in such a way that, when they are radiated from two loud-speakers in front of the listener, the signal from the dummy's right ear will only go to the listener's right ear and that from the dummy's left ear only to the listener's left ear, just as in earphone listening but with the proper free-field coupling. But, in fact, the following situation may arise under experimental conditions: The sound from each loudspeaker

---

[1]The anechoic room is designed to absorb as much sound as possible (see Fig. 2.4). The walls consist of a number of baffles with highly absorptive material. Any fraction of sound reflected is directed towards another baffle instead of back into the room. Thus, any echo is almost completely suppressed.

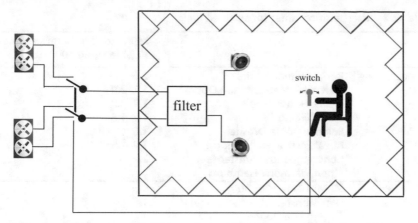

Fig. 12.8 Setup for the listening and subjective evaluation phase in an anechoic chamber. Two concert halls are compared at a time, the listener being allowed to switch back and forth between the two recordings

goes into both ears and not just to the "near" ear of the listener, as would be desired. In other words: there is "cross-talk" from each loud-speaker to the "far" ear. Ways have been developed to cancel out this unwanted effect by filtered compensation signals from the loudspeakers. Finally, a preference matrix could be established whose entries indicate how many times each hall was preferred by each listener.

In the second phase of this study the correlation with objective parameters was considered. The parameters include volume and width of the hall, time delay between the direct sound and the first reflection at the listener's seat, the reverberation time T_r, interaural coherence, and energy from measured impulse response of the dummy's ears with a powerful spark source on the stage. Important findings are that the greater the reverberation time T_r, the greater the preference for these halls, but that for T_r above 2 s the consensus preference is smaller. The width of the halls is suggested to have a subjective significance via a substantial correlation with T_r for many of the halls under study. If the volume of the hall becomes larger, the correlation with consensus preference is mostly negative. One may say that once a hall has reached a certain size, don't make it any bigger—or be prepared to suffer acoustically (Table 12.2).

Table 12.2 Reverberation times for selected halls

		Selected reverberation times T_r (s) (around 1000 Hz)
Concert halls	Berlin Philharmonics	2.0
	New York Metropolitan Opera	1.7–2.0
	Opera House Oslo	1.7–2.0
	Scala Milano	1.6–1.8
	Semper Opera Dresden	1.6
	Royal Festival Hall London	1.4–1.5
	Concertgebouw Amsterdam	2.2
	Elbphilharmony Hamburg	?
Churches	St. Michaelis Hamburg	6.3
	Ulm Münster	12
	Cologne Cathedral	13
Others	Living room with furnitures	0.5–0.6
	Audio studio	0.2–0.3
	Anechoic chamber (counterpart of reverbation chamber)	0.01

12.4.2   Absorption Coefficient α

Acoustic absorption refers to the process by which a material, structure, or object takes in sound energy, as opposed to reflecting the energy. Part of the absorbed energy is transformed into heat and part is transmitted through the absorbing body. Materials for sound absorption can serve to reduce reverberation times, suppress disturbing echoes or lower the noise level in performance rooms. The absorption coefficient α describes the portion of absorbed/ impinging sound energy:

$\alpha = 1$: no sound reflection;

$\alpha = 0$: full sound reflection.

α depends on the angle of incidence.

To determine the value of α for an acoustic absorber one can place an extended probe of the material under investigation in a large reverberation chamber with "sound-hard" walls, having reverberation times of $T_r = 10$ s and more, when empty. With the probe T_r will diminish. Then α is calculated according to Eq. 12.8. Other measuring procedures are the impedance tube (only for vertically impinging acoustic signals) and the laser-vibrometer.

The following table (Table 12.3) provides some values for α of common materials.

In this context it is interesting to note that the audience in a hall can be considered as a high frequency absorber, absorbing all sound contributions

Table 12.3 Absorption coefficient α

Material	Absorption coefficient α at		
	125 Hz	500 Hz	2000 Hz
Open window	1.0	1.0	1.0
Closed window	0.1	0.02	0.03
Plastered stone wall	0.01	0.02	0.03
Wood panelling wall	0.1	0.1	0.1
Carpet (medium thickness)	0.1	0.2	0.35
Curtains (with folds)	0.1	0.3	0.5
1 wooden chair	0.01	0.02	0.05
1 upholstered chair	0.2	0.3	0.35
1 listener on a upholstered chair	0.2	0.5	0.6
Well attended hall		0.95	

from approximately 500 Hz on upwards. In this, the surface occupied by the audience is a determining factor, while the seating density is of minor significance. The same number of persons affects a higher degree of absorption when distributed over a larger surface.

The role of curtains is particularly interesting, because they can be used without structural changes. As a porous material, they absorb, as do carpets, preferentially high-frequency contributions. They should be hung with folds, and when suspended at certain distances from the wall, lower frequencies will be included. Among the large number of designs for efficient acoustic absorbers, we select three that are frequently used.

Porous material such as textile, glass or mineral fibers are good sound absorbers because of the internal flow resistance leading to partial annihilation of acoustic waves. Wedges made of such material are readily used for coating anechoic rooms, for instance. Any reflection still prevailing may be avoided by shaping the wall in form of adjacent pyramids or wedges. The absorbing effect is always acting in a broad frequency band (Fig. 12.9a).

The plate absorber consists of a paneling, often made of wood, which is placed in front of a wall. The air cushion in between acts like a spring and forms together with the mass of the plate an oscillating system. (Fig. 12.9b). A sound wave impinging on the plate excites the system to a forced vibration. The vibration is damped by internal losses in the plate material and by porous absorbers, if inserted into the air gap. There is a resonant frequency which can be adapted to the current needs and where the strongest absorption occurs. It usually lies between 70 and 300 Hz.

A plate absorber perforated with holes in front of a stiff wall can be considered as a large number of parallel Helmholtz resonators. Such a resonator is

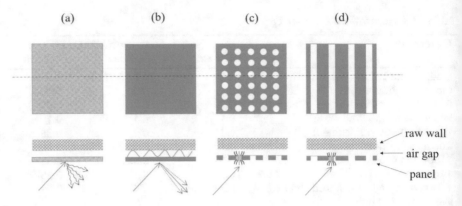

Fig. 12.9 Several examples of absorbers: **a** porous material, **b** plate absorber (wood paneling), **c** perforated panel of the Helmholtz-type (with air flow through a hole), **d** slotted panel. The air gap can be (partially) filled with porous material

a cavity with a small opening (see Fig. 12.9c). If one puts a panel with many holes in front of a wall, the air volume between plate and wall acts like a spring, just as for the plate absorber. Now, the vibrating mass is contained in the holes and the resonators are damped by the flow resistance in the openings. The resonance frequency (typically higher than 600 Hz) is determined by number and size of the holes as well as the thickness of the air cushion.

There are many more types of sound absorbers, e.g. the slotted panel shown in Fig. 12.9d.

12.5 Geometrical Room Acoustics

Having discussed some aspects of wave acoustics and its statistical properties, we now turn to situations, where the wavelength is very short in comparison to the room dimensions. Then, the geometrical approach of sound beams is appropriate, which propagate along straight lines and are reflected by the walls. The listener is reached by different acoustic beams: there is the direct acoustic beam from the sound source, and after that, beams which have been reflected one or multiple times at walls or the ceiling (compare the schematics of Fig. 12.6). Single strong returns within about 50 ms are perceived as augmentation of sound intensity, those after 50 ms as echo, which is disturbing.

When planning the acoustics of a hall, it is therefore advisable to study in a model the distribution of the first returns. The propagation of sound beams can be conveniently traced with light beams. The geometry of beam reflections in two dimensions is shown in the examples of Fig. 12.10 for a central source

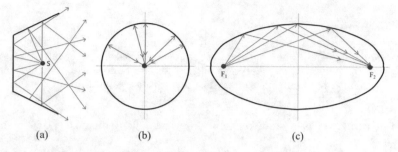

(a) (b) (c)

Fig. 12.10 Geometrical acoustic picture of sound reflections in a model with light beams: reflection with **a** planar walls, **b** half-circle wall, **c** elliptical wall

S with (a) a set of planar reflecting walls, (b) a half-circular reflecting wall and (c) a half of an elliptic wall. In general, sound signals arriving after short time intervals at the site of listeners are integrated in loudness. In the case (a), there are several sites where the sound is louder than the surroundings.

When the wall is a half circle and the sound source is at the center of the circle (Fig. 12.10b), all sound beams return back to the place of the sound source. When the wall is a half ellipse and the sound source is set at one of the focal points, then all sound beams reach the other focal point (Fig. 12.10c).

We can readily place these scenarios into 3D space. Just imagine angled walls of a hall (a), a half sphere (b), and the upper half of a prolate ellipsoid (c).

A quick digression: If the sound source emits strong ultrasound waves, the sound source itself is destroyed in the case of (b), and an object (e.g., an expensive wine glass or a crystal) set at the other focal point is damaged in the case of (c).

A "flutter echo" is produced by sound traveling quickly between two parallel reflective surfaces. In a room, depending on its size, the nature of this echo can vary from a fast procession of separate, distinct sound events to a seemingly unbroken series of echoes. Such a "flutter" or "chatter" effect is normally not desirable. It can be avoided just by introducing a small tilt angle between the surfaces.

Intermezzo

If the reflecting boundary is not flat (with edges) but curved, the reflected sound is kept there. In fact, the impinging waves always remain close to the wall. This is the origin of the "whispering gallery-waves" which have been heard in many examples of circular, hemispherical or elliptical enclosures. Famous examples include St. Paul's Cathedral in London, St. Peters in Rome, the Cathedral of Brasilia in Brazil, the Alhambra in Granada, Spain, and many more. Sir C. V. Raman, when visiting London in 1921, studied how sound travels along the gallery of St. Paul's Cathedral.

Let us be impressed by the mausoleum Gol Gumbaz in Bijapur, half-way between Mumbai and Bangladore, India, with its huge dome (the left figure below) [5]. At a significant altitude a gallery was constructed encompassing this dome at a certain level (the right figure). For people wandering along this gallery, any acoustical sound uttered would be heard very clearly at any other spot along the path of the gallery.

Left: Gol Gumbaz Mausoleum (public domain by Ashwatham); right: view of the gallery (CC BY-SA 3.0 by Mukul Banerjee)

The acoustic waves guided along the curved wall are related to Rayleigh waves moving directly at the boundary between wall and air. This kind of wave may occur at seismological events on the surface of the earth or on water surfaces. These are purely two-dimensional waves at interfaces, but their impact for the discussed whispering effect remains under discussion.

In his article "Raumakustik (Room acoustics)" Vern O. Knudsen discusses instructive examples on reverberation and absorption as investigated in large halls [6]. In the Royal Albert Hall, London, with a very elevated dome-shaped top, strong echoes were originally heard. These echoes were much reduced by spanning a baldachin of heavy cloth at the ceiling. Ceiling structure turns out to be always of major importance for concert hall design (compare Fig. 12.4).

The Mormon Temple in Salt Lake City, Utah, with up to 2500 visitors has an optimal reverberation time and a generally excellent acoustics. An important factor is the adaptation of the ceiling to improving the acoustic properties. One has added animal hairs to the plaster for the ceiling (!). We find the structural outline of this cathedral in Fig. 12.11.

Some tests for the acoustical properties of this immense hall, especially reverberation times, were done with pistol shots. In Fig. 12.11 the locations of the pistols and microphones are indicated. One finds for the transverse direction response curves with a sequence of small echoes, whereas for the pistol shot at the location of the pulpit one measures further down a smooth reverberation curve. But such a shot is not dangerous, as in the novella "The Shot from the Pulpit" by Conrad Ferdinand Meyer (1825–1898), a German poet. It just serves to create a short but intense acoustical signal (with a SPL of well above 100 dB), subsequently analyzed via a microphone. (No gunfire (!), due to superheated gases exiting the muzzle or whip-like "snaps" caused by the sonic boom as a projectile moves through the air at supersonic speeds.)

Finally, as to the interaction of music and room, we note that:

- Only in rooms with excellent acoustic properties can individual instruments merge to a global sound and develop spatial greatness.

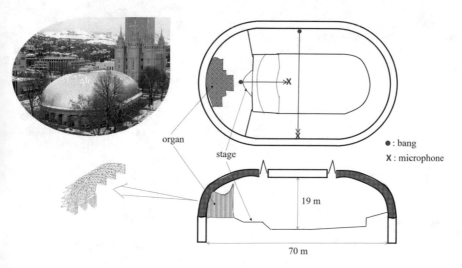

Fig. 12.11 Left: Tabernacle on Temple Square in Salt Lake City, Utah, U.S.A. at Christmas time (CC-BY-SA 3.0 by Leon7), right: the structural outline of the cathedral (illustrated according to [6])

- There has to be a balance between stage and auditorium.
- Intensity relations between instrumental groups have to be equal throughout.
- Reflections, reverberation and their regulation have to be evaluated.
- Differences between empty and occupied halls should not be too large.
- Musicians should be able to hear themselves and others.
- A correct arrangement of instruments according to their radiation characteristics has basic importance.

References

1. D. Lubman, Archaeological acoustic study of chirped echo from the Mayan pyramid at Chichén Itzá. J. Acoust. Soc. Am. **104**, 1763 (1998)
2. E. Meyer, E.-G. Neumann, *Physikalische und Technische Akustik* (Friedrich Vieweg und Sohn, Braunschweig, 1974), pp. 52–85
3. W.C. Sabine, *Collected Papers on Acoustics* (Dover Publications, New York, 1964)
4. M.R. Schroeder, D. Gottlob, K.F. Siebrasse, Comparative study of European concert halls: correlation of subjective preference with geometric and acoustic parameters. J. Acoust. Soc. Am. **56**, 1195–2001 (1974)
5. https://en.wikipedia.org/wiki/Gol$_-$Gumbaz
6. V.O. Knudsen, Architectural acoustics. Sci. Am. **209**, 78–95 (1963)

Part VI
Music, Mind and Society

.... *Der Vorhang zu*
und alle Fragen offen.

.... *The curtain closed*
and all questions stay open.

—*Bertolt Brecht*

13

Music Analysis with Phase Portraits

13.1 Introduction

We, the authors, are neither psychologists nor philosophers. This we have to say clearly at the beginning of this chapter. Nevertheless, we will discuss some relationships between music and our feelings/emotions via our methods of mathematical/physical analysis.

During the past centuries many musical pieces have been analyzed in various ways: analyses based on the theory of harmony and scales, as well as rhythms, on historical and cultural backgrounds, on comparison with other pieces, and so on. These are, if we may say, analyses on the side of hardware of musical pieces.

What would then be the software side of musical pieces? We think that it is something that interacts with our emotions. Some musical pieces sound so sad to remember your dead friend, some make you happy, even to go up to heaven, some encourage you to decide to choose a difficult way. In general, pieces with major scales give you a discrete and sure feeling, while those with minor scales make you unsure and unfinished. But this is not all and not so simple. One musical piece may give you a feeling of happiness which you like to share with many people, while a happy feeling from another piece you like to keep for yourself, for example. What causes such differences?

Can we develop some tools to analyze such features of music? Can we find characteristic patterns in musical pieces for a sunny summer day or a storm

---

The original version of this chapter was revised. The second line of equation 13.1 on p. 364 of this chapter has been changed from "$y_p = y \cdot \sin\alpha + z \cdot \cos\alpha$" to the correction version "$y_p = y \cdot \cos\alpha + z \cdot \sin\alpha$". The correction to this chapter can also be found at DOI https://doi.org/10.1007/978-3-030-68676-5_15.

in winter? How did Antonio Vivaldi manage to express different scenes in his "Four Seasons"?

Listening with closed eyes to Chopin's Etude "Winter wind" Op. 25 No. 11, in A minor played by Maurizio Pollini, KT feels silver powder intermittently showering on her. Such feelings can be very personal, but some people agree with her. Maybe there are some common feelings. How can we express them?

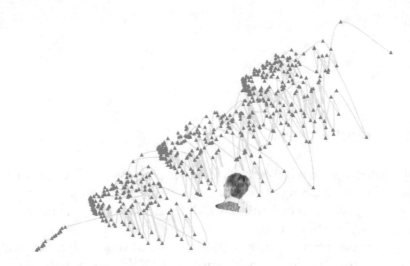

Fig. 13.1 Silver powder falling on the listener

When KT tries to explain her "silver falling" feeling, she may show a scene like in Fig. 13.1. We are sure that after seeing this image people understand better what she wishes to convey. So, should we draw images which you perceive during listening to music? Yes. Drawing or painting is a good way to visualize the images. However, drawing or painting is as subjective as the image in your mind. Can we form the image more objectively?

Actually, this figure was not drawn by hand, but plotted as a 3D projection of a phase portrait (the frequency versus the velocity of frequency change) along a time axis.

This method was presented for the first time by KT in order to put her feeling (walking down a spiral staircase with herself spinning) in visual form, when playing Johann Sebastian Bach's C major prelude of "The Well-Tempered Clavier, Book I" [1]. In this chapter we briefly explain phase plots which are often used in nonlinear physics, and how we can apply them to musical pieces. With help of some examples we discuss whether this method can serve as a tool to analyze the software-side of music.

Furthermore, we will demonstrate how music scales look like.

13.2 Trials with Phase Plots

13.2.1 What Is a Phase Plot?

A phase plot may represent the dynamics of location (x) and momentum (proportional to $y = dx/dt$) of a mass-spring system. Instead of showing directly the time evolution of x and y (by sine and cosine functions) it presents the motion on a xy-plane by a closed trajectory (circle or ellipse), thus providing a compact image of the system's oscillatory dynamics.

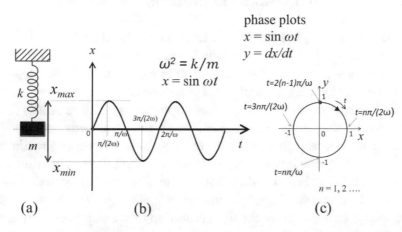

Fig. 13.2 a Mass-spring system, **b** time course of the harmonic oscillation (it continues indefinitely towards the right direction), **c** phase plot of the time evolution. $t = 0$ (red point) is defined as the time when the weight comes back to the original equilibrium position after having been pulled down for the first time

The simple example of oscillations of a spring is shown in Fig. 13.2. A weight is attached to one end of the spring, and the other end is fixed at the ceiling. If you leave it as it is, it stays as it is (equilibrium state). If the weight is pulled down, the spring tries to recover its original state. However, usually an "overshoot" occurs and the weight starts to oscillate vertically. The time course of the height of the weight shows the form of a sine curve. If no energy dissipation occurs, this oscillation lasts forever. It means that the curve of the time course of oscillation is indefinitely long. If energy is dissipated, the amplitude of the oscillation decreases and finally the weight stops moving. Even in such a case, the curve of the time course can be very long. Figure 13.2c illustrates a plot of the location (x) of the weight versus its velocity ($y = dx/dt$) for the case of zero energy dissipation. As shown in this figure, a phase plot presents the oscillation in a very compact manner, instead of having to show a long, long sine curve [2].

If the energy is dissipated by friction during oscillation, the motion is damped, and the closed curve of the phase portrait turns into an inward curling spiral, which asymptotically ends in the origin, acting here as a point attractor. When the energy is dissipated fast, the curve approximates the origin also fast. When the energy is dissipated slower, it takes a longer way to approach the origin. So one can see how the motion is damped by looking at the trajectory.

13.2.2   Three-Dimensional Expression

Since the phase plot gives no explicit information for the time evolution, we never know when the point (x, y) which you see now corresponds to the real movement with time. If the time course is not essential, in physics this aspect is advantageous for analysis of physical phenomena.

But when the time evolution becomes important, an additional axis (z-axis) vertical to the xy-plane is introduced, which pulls the trajectory along the z-axis. If your view is perpendicular to the xy-plane, one sees exactly the same phase plot as without introducing the z-axis, similar to the situation where you look at a helix-shaped spring from its head (or tail) along its axis—you see only a circle, and you have information of neither the length nor the pitch of the spring. In order to see the time evolution of the trajectory, you have to change the angle of view.

Mathematically we can "change the angle of view": We rotate the yz-plane around the x axis and the xz-plane around the y axis, and plot the projection on a new $x_p y_p$-plane. We use the following equations to plot the projection (x_p, y_p), when at first the object is rotated α radians around the x-axis and then β radians around the y-axis:

$$
\begin{aligned}
x_p &= x \cdot \cos \beta - (-y \cdot \sin \alpha + z \cdot \cos \alpha) \cdot \sin \beta, \\
y_p &= y \cdot \cos \alpha + z \cdot \sin \alpha
\end{aligned} \tag{13.1}
$$

Figure 13.3 shows a 3D presentation of the phase portrait of (a) a harmonic oscillation and (b) a damped oscillation (for the case with energy dissipation). Although this kind of 3D presentation does not add any new physical information, there is the advantage to see at a glance what is going on with time.

The coordinate z is arbitrary with respect to x and y, but a proper value should be found by trial and error. If the value is too large (very large expansion), the 3D projection is almost linear and you do not see any characteristic shape. If a too small value is taken (not enough expansion), the 3D projection is too dense, and you do not see the shape either. If the value for z would be a factor 10 larger than in Fig. 13.3a, the 3D projection were almost linear.

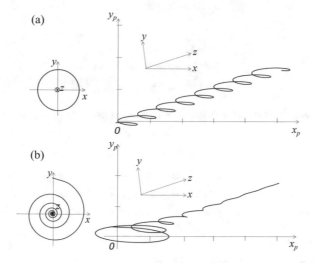

Fig. 13.3 3D plot of the phase portrait of **a** a harmonic oscillation and **b** a damped oscillation

13.2.3   Applying Phase Plots to Musical Pieces

This dynamical approach is now used for our analysis. For the music we can attribute x to the frequency f and y to the change of the frequency per unit time df/dt. The frequency corresponds to the pitch of each note. Different from an oscillator like a physical pendulum, however, x and y are not continuous, but assume discrete values. Therefore, we define:

$$x_n = (f_n + f_{n+1})/2$$
$$y_n = (f_{n+1} - f_n)/m_n = \Delta f_n / \Delta \tau_n, \qquad (13.2)$$

where f_n is the corresponding frequency of the nth note, $\Delta f_n = (f_{n+1} - f_n)$, and $m_n (= \Delta \tau_n)$ is the time interval between the nth and $(n+1)$th note (see Fig. 13.4). For example, if the quarter note is set as $m = 1$, then $m = 4$ for a whole note, $m = 2$ for a half note, $m = 1/2$ for an eighth note, and further. When a piece consists of many 16th notes, it is convenient to set the 16th note at $m = 1$. Then, $m = 16$ stands for a whole note.

Since we have only discrete points, we have to connect them. After some trial we decided to use a polynomial approximation. There is no physical or

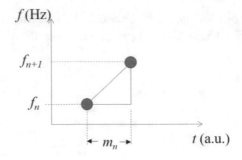

Fig. 13.4 Two adjacent notes, nth and (n+1)th, with pitch f_n and f_{n+1} at time interval m_n

mathematical reason for our decision, though. Just sharp edges are avoided, because they do not fit to the musical sound. For instance, Bach's notes create a continuous curve of sound, where subsequent notes are intimately connected.

13.3   Analysis—Some Examples

13.3.1   Bach and Gounod

Bach's prelude 1 in C major from The Well-Tempered Clavier Book 1 is known by many music friends and can be played even by a beginner for learning piano. This is a brilliant piece: it consists only of broken arpeggios and still leaves a great impression on people. Various chords are arranged in an order such that it conveys a feeling to get wide and fine, get narrow and nervous, recover to be fine, again become uneasy, recover from it, …and finally come to the goal safe and sound. When I play this piece, I feel as if I am going down a spiral staircase, spinning around myself on every stair.

When I sing Gounod's "Ave Maria" on the other hand, I feel as if I am flying up to the sky accompanied by angels who are floating and spinning. Although the piano part is "almost" identical, I go down with Bach to the steady ground, and I go up to the sky with Gounod.

Charles Gounod had the ingenious idea to use Bach's ingenious piece as a piano accompaniment of a song. He improvised the beautiful melody and Pierre-Joseph-Guillaume Zimmermann (French composer and pianist, also the father in law of Gounod [3]) transcribed it. In 1853 an arrangement for violin with piano was published as "Meditation". A little later "Ave Maria" in Latin was published.

In a previous paper we tried to present my feeling (going down the staircase with turning myself) visually, by using phase portraits [1]. In this section we will compare Bach's prelude and Gounod's Ave Maria, and try to find out why I am flying up with Gounod.

If you listen to Bach's prelude and Gounod's Ave Maria, the piano parts sound very similar. However, there are three differences. The first difference is obvious: Gounod used the first 4 bars additionally as introductory part and repeated the main part twice before the coda.

The second difference is almost impossible to notice by listening. Even for some professional pianists it is a surprise to find this difference.

Bach's wonderful piece was once slightly modified by C. F. G. Schwencke [4] and published in 1801. He added one bar between bars 22 and 23, because he thought that it is not good to have two diminished seventh chords in sequence (see Fig. 13.5). He probably had the suspicion that it was Bach's mistake. But after adding the new bar, the piece was not improved at all: an interesting tension and release in Bach's original was lost. Nowadays Schwencke's version is not played so often. A notable exception is the "Ave Maria" (or "Meditation") of Gounod.

Fig. 13.5 Additional bar between the bars 22 and 23 inserted by Schwenke

The third difference is shown in Fig. 13.6. The first and the second long notes (and those in the repetitive part, the 9th and 10th notes) in each bar of Bach's original (Fig. 13.6a) are replaced by sixteenth notes in Gounod's Meditation (Fig. 13.6b). This difference is often not noticed or ignored by beginners or some people who use the pedal all the time.

Let us think what this difference means. In Gounod's case the eight successive notes at the beginning of each bar are pure arpeggios, while in Bach' s case the first note (a half note) affects the following seven notes, and the second note (a double dotted quarter note) affects the following six notes as members of a chord. It means that there is a "basso continuo" in Bach's piece, but not in Gounod's composition. The pitch of the basso continuo is C4 at the beginning. It decreases time by time in a back and forth manner, and finishes with C2. The basso continuo plays the role of a backbone throughout the piece and

Fig. 13.6 Structural difference of **a** Bach's original and **b** Gounod's modification

the second long note supports the backbone. Instead of the backbone and its support there is a melody in Gounod's piece.

In order to distinguish the three kinds of notes in Bach's work (a half note, a double dotted quarter note and a sixteenth note) in our analysis explicitly, we separate each group of notes into three categories (see Fig. 13.7) [1]:

(a) 8 sixteenth notes (the first two longer notes are considered sixteenth notes, as well),
(b) 1 half note (the lowest pitch),
(c) 1 double dotted quarter note (the second lowest pitch) in combination with the lowest note.

Fig. 13.7 Examples of categories **(a)**, **(b)** and **(c)**. Red arrows indicate three roles of the lowest tone and blue arrows two roles of the second lowest tone

The lowest note has three roles for each half bar, as indicated with red arrows in Fig. 13.7. In a similar way the second lowest note has two roles, as indicated with blue arrows in Fig. 13.7. Categories (a), (b) and (c) are plotted independently. For Gounod's Meditation, there is only the category (a) for the piano part, and therefore, the category (a) and the violin part are plotted independently.

In Fig. 13.8, 3D presentations of the phase portrait for Bach's prelude and Gounod's Meditation are shown. For Bach's prelude the "backbone" or the "staircase" (thick brown curve in Fig. 13.8a) and its "support" (green curve) stick to the blue curve and go towards the lower pitch.

Such a backbone and its support do not appear in Meditation (Fig. 13.8b). The red curve stands for the melody of violin and the blue curve for the piano.

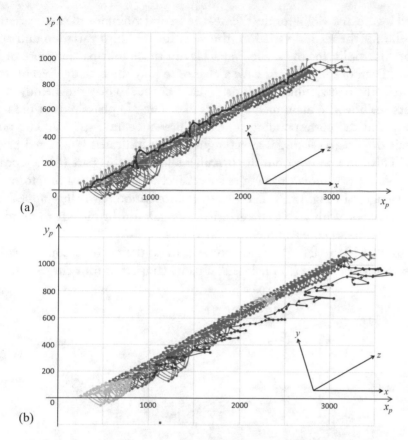

Fig. 13.8 3D Phase plots of **a** Bach's prelude in C major: blue, dark brown and green curves correspond to the curve for category A, B and C, respectively, **b** Gounod Meditation, blue curve: piano (category A of Bach's original), yellow part at the beginning: added by Gounod, yellow part in the middle: added by Schwencke, red curve: melody

Gounod's melody moves away from the piano time by time towards the higher pitch.

This could be an explanation: at hearing (or playing) Bach's prelude one goes down along the staircase, and at hearing (or playing, singing) Gounod's Ave Maria one flies up toward the sky.

13.3.2 Haydn: String Quartet "Emperor"

The second movement of Haydn's "Emperor" quartet (the string quartet in C major Op. 76 No.3) is well known by its melody as the German national hymn. This piece consists of a theme and four variations. The main melody is played by the first violin for the theme, the second violin for the first variation, the cello for the second variation, the viola for the third variation and again by the first violin for the finale. Papa Haydn treats four parties more or less equally. Each of four persons have a chance to play the melody. Exception is, of course, the first violinist. He (or she) has a chance to play the melody twice.

Let's see how the main melody looks like. The 3D phase plot of the main melody and the corresponding notes are presented in Fig. 13.9. The phase portrait of part (A) includes a rectangular form indicated by the red arrows (left). This form comes from the ornamental. Since this part (A) is repeated twice, we see two similar rectangles successively. Part (C) has a ring form (see arrows (right) in Fig. 13.9). Because of the repetition, two rings appear one after the other. With such a characteristic portrait for the melody in mind, we will analyze the theme and the four variations.

Figure 13.10 presents the phase portraits of (a) theme, (b) second variation, (c) third variation and (d) finale. Each panel consists of four curves: blue for

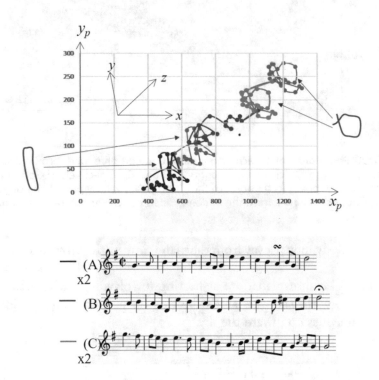

Fig. 13.9 Phase plots of the main melody of Haydn's string quartet Emperor, **A** the first and the second phrases (repetition not shown), **B** the third phrase, **C** the fourth and the fifth phrases (repetition not shown). Red arrows indicate the characteristic forms

the first violin, green for the second violin, red for the viola and purple for the cello. (About the first variation we will talk later.)

In the theme (Fig. 13.10a) the first violin plays the melody. Therefore, the characteristic curve with two rectangles and two circles is in blue (indicated by arrows). Here you can see that the other three curves (green, red and purple) are on the left side (lower pitch) of the blue curve. It means that second violin, viola and cello support the first violin with lower frequencies. The cello plays the role of a stable base at the bottom. The viola and the second violin fill the space between the base and melody.

In the second variation the cello plays the melody one octave lower (Fig. 13.10b). The portrait of the cello is smaller because of the low frequencies, but nevertheless it shows the characteristic curve with two rectangles and two circles, indicated by arrows. Here, the cello is the base and at the same time the "star". The cello is sovereign. The viola and the second violin obey faithfully the way of the cello: three curves are close together. Probably, only with these three, music would be too plain and rather poor. The first violin fills such vacancy with higher pitches and faster movement so that the music is more interesting for listeners.

Next comes the turn of the viola. In the third variation (Fig. 13.10c) the viola plays the melody (see red rectangles and rings), sandwiched between the first violin from the higher frequencies and the second violin from the lower frequencies. When the first violin plays with high pitch and high speed (see the large triangular shape of the blue line), the cello starts playing as an anchor in order to avoid "flipping out" of the first violin.

The first violin plays the melody in the finale (Fig. 13.10d), too. From the repetition of (A) on, the pitches are one octave higher than before, so that the second rectangle and the following two circles are much larger. Correspondingly, the other three also play with higher pitches and larger movements. Nevertheless, the base with the cello remains very stable.

So far everything sounds OK. The first violinist likes to expose himself. He (or she) is the best of four, he (she) thinks. The cellist knows how important he (she) is: without him (her) the whole concept does not work (Germans say "Alles kaputt!"). The violist is usually very modest. He (she) is happy, if he (she) has a little chance to be the main actor. He (she) does not complain to have such an occasion very seldom. The second violinist often tends to be dependent on the first violinist (Heimeran used the term "Schwester-Geige" [5], whether ever he (she) likes it or not). With these four different and even difficult characters they can play the "Emperor" more or less together and even harmonically except for the variation 1.

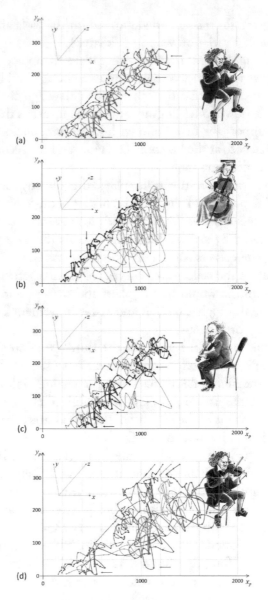

Fig. 13.10 Phase plots of Haydn string quartet Emperor, **a** theme, **b** second variation, **c** third variation, **d** finale. Blue: first violin, green: second violin, red: viola, purple: cello (illustrations by M. Covarrubias)

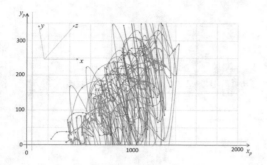

Fig. 13.11 Phase plots of the first variation of Haydn's string quartet Emperor. Note that the plots for the first violin run over the panel

The first variation is played only by the first and second violin. The second violin plays the melody, as shown in the green curve in Fig. 13.11. If you look at the figure carefully, you will recognize two rectangles and two circles representing the melody. The awful noise-like curves in blue are the part of the portrait for the first violin: note that not all are seen, because the field of the portrait for the first violin is much larger.

The whole portrait of the first violin is presented in Fig. 13.12a. It is extremely wide in both directions x_p and y_p. The corresponding plot for the second violin is seen in Fig. 13.12b. The second violin is totally wrapped by the first violin. The melody is wrapped by super-ornaments. Do they know this? The viola and cello are relieved not to be involved in the problem between the first and the second violins.

Figure 13.12c is the case 1 when the first violin says, "I am the greatest". He (she) is interested only in showing himself (herself), standing at the center point and believing that he (she) is the best. He (she) does not know how bad the music is. The audience is not happy, because they cannot hear the melody, they hear only hectic up and down.

If the first violinist is really good, he (she) understands the intention of Papa Haydn. The filigreed second violin should be protected with a wrap. This wrap could be the cocoon of a silkworm to protect a living creature inside. The excellent first violinist produces a transparent cocoon. He (she) covers the second violinist completely, but nevertheless you can hear the melody which the second violinist plays. This is the case 2 of Fig. 13.12d.

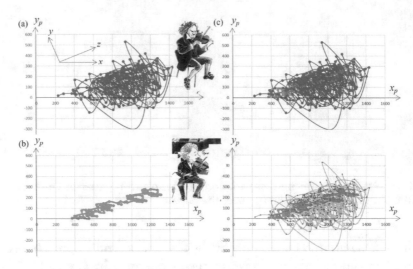

Fig. 13.12 Phase plots of the first variation of Haydn string quartet Emperor, **a** first violin, **b** second violin, **c** playing together (case 1), **d** playing together (case 2) (illustrations by M. Covarrubias)

Coffee Break

The worst thing for women is to get married with a male first violinist. The worst thing for men is to get married with a female first violinist. The worst thing for the world peace is a marriage of the male and female first violinists.

13.3.3 Rossini: Guillaume Tell Overture

Now we consider an orchestra score. KT can read musical notes more or less. When she sees musical notes for a song, she can hum its tune. However, to understand orchestra scores is not so simple. Many instruments play with different pitches and different rhythms at the same time. She can read a line of score for the first violin, and she can guess the melody. But it is difficult to know what other instruments are doing. Different from laymen like us, professional conductors can read scores as a whole. They know when a percussion person should raise his hand for the next action.

Here we use as an example a very popular piece which most people know: the last part of the overture of Guillaume Tell by Rossini, starting with a trumpet fanfare. The bars 1–8 (with upbeat) are for the fanfare of trumpets, followed by its echo by horns. The next bars (bars 9–17) are only rhythms (ta ta - tatta - ta -ta - tatta - ta -ta- tatta - ta -ta- tatta - ta -ta -taaa) of trumpets, horns and timpani with different pitches.

From bar 18 to 25, more instruments join to play: all strings, the first clarinet, the bassoons, the third and fourth horns and timpani, but without trumpets and the first and second horns. Among them the first violin and the first clarinet play the melody, and other instruments contribute in rhythm as a background.

From bar 26 on further instruments are involved: piccolo, flute, the first and the second oboes, the first horn, the trumpets, and the trombones. Here (bars 26–33) piccolo, flute, the first oboe, the clarinets and the first violin play the melody: piccolo—the highest pitch, flute, the first oboe and the first clarinet—middle pitch, the second clarinet and the first violin—the lowest pitch. Other instruments play rhythm with various pitches according to the instruments.

Between bars 34 and 41 piccolo (the highest), flute and the first oboe (the second highest), clarinet (the third) and the first violin (the lowest) play the melody. Others play rhythm. Triangle joins here. The bars from 42 to 49 are the repetition of bars from 34 and 41.

Figure 13.13 shows a 3D plot of these 49 bars (with an upbeat). The scores for each instrument are plotted independently: the first violin, the second violin, viola, cello, flute, piccolo, clarinet-1, clarinet-2, bassoon-1, bassoon-2, horn-1, horn-2, horn-3, horn-4, trumpet-1, trumpet-2, trombone-1, trombone-2 and timpani. Since the notes of the double bass in these 49 bars are exactly the same as those of the cello, they are not plotted extra.

If you look at Fig. 13.13, you can understand what Rossini expresses in the score. With the fanfare (bars 1–8) horses come out and line up (bars 9–17). After a short silent moment, they begin to gallop straight (bars 18–25). Then, more horses join and the sound of gallop becomes louder (bars 26–33). Horses are running and running with swinging their manes up and down (bars 34–49).

I am one of these horses. I am running and running straight toward the goal. I am trying hard not to drop behind. I have to run faster, faster, ...

A conductor's role is to let these horses run to the goal all together.

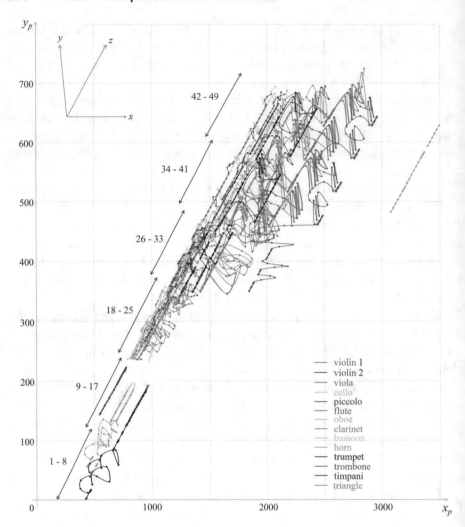

Fig. 13.13 3D presentation of the phase portrait of Guillaume Tell overture. The coordinates at the upper left indicate the projection of directions of x (frequency), y (frequency change) and z (time course). The numbers (1–8) mean the bars from the 1st to the 8th after starting the fanfare, (9–17) from the 9th to the 17th, and so on

13.4   Comparison of Various Scales

As a further application of the phase portrait, we use it for comparing various tonal scales. We take simple parameters for all plots in 2D (f vs. df/dt): constant time interval, fundamental tone C3, tone interval from C3 to C5.

13.4.1   Heptatonic Scales

Among the heptatonic scales we have the church modes (including major and minor scales) as introduced in Sect. 3.2.4, Gypsy scales of the Sinti and Roma (5.2.1), misheberach scale of the Jewish music (5.2.2), ragas of Indian music (5.2.3), the maqām of Arabian music (5.2.4), and others.

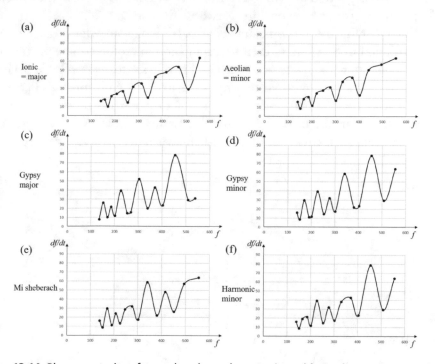

Fig. 13.14 Phase portraits of two church modes **a** Ionic and **b** Aeolian, **c** Gypsy major scale, **d** Gypsy minor scale, **e** misheberach and **f** harmonic minor scale with the base tone C

Phase plots of the two church modes (Fig. 13.14a Ionic and b Aeolian) have in principle the same shape. Only their starting point is shifted. This is true for other church modes, too: Their phase portraits are identical, if they are laterally shifted. This suggests that all church modes belong to one family. On the other hand, two Gypsy scales (Fig. 13.14c major and d minor) have forms analogous to each other, but completely different from church modes. Portraits of the Jewish scale misheberach (Fig. 13.14e) and the harmonic minor scale (Fig. 13.14f) are not far from those of Gypsy scales.

It is easy to understand at a glance on Fig. 13.14 that Gypsy and misheberach scales belong to families very close to each other, but are far from the family of the church modes. However, since the harmonic minor scale has been developed from Aeolian, it plays the role of a bridge between the church mode family and the misheberach family (refer Sect. 5.2).

If we proceed to the Indian ragas, some scales are identical to church modes and a Gypsy mode and others are completely different. As seen in Fig. 13.15, kaphia is exactly the same as the Doric, but marava, purvi and todi have completely different portraits, which can not be attributed to any other shapes. For our ears they sound like "fu nya fu nya fu nya fu nya ......" with the interval of a perfect fourth or a major third between "fu" and "nya".

Phase portraits of Arabic maqām (Fig. 13.16) look somewhat strange to us. As mentioned in Sect. 5.2.4, their scales include microtones. The portrait of "rast" (a) is very flat and smooth because of two microtones (see Table 5.1). It sounds like "nyu-nyu-nyu-nyu-nyu-nyu-nyu-nyu-..." with intervals between a minor second and a major second. The other three scales possess one jump in each octave on the base of a small "fluctuating" curve. A typical melody imagined from such curves is "nyu-nyu-nyu-nyu-**nya**-nyu-nyu-**nya**-nyu...".

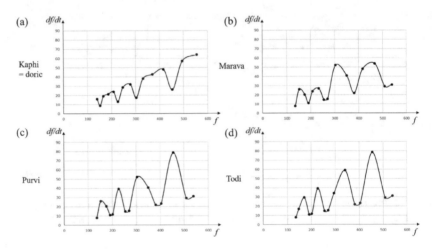

Fig. 13.15 Phase portraits of raga scales: a kaphi—exactly the same as doric), **b** marava, **c** purvi, **d** todi

13.4.2   Pentatonic Scales

As pentatonic scales we introduced European examples (3.2.2), Chinese scales (5.4.1) and Japanese scales (5.4.2)).

The phase portraits of the European pentatonic scales (major variant and minor variant), all Chinese traditional scales and some Japanese scales have a comparable form when considering lateral shifts (see Fig. 13.17a, b and c).

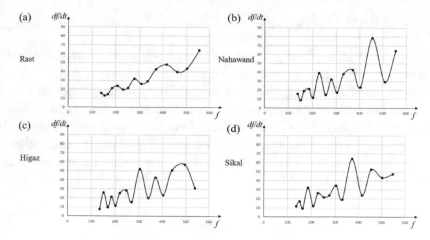

Fig. 13.16 Phase portraits of maqām scales: **a** rast, **b** nahawand, **c** higāz, **d** sikal

This indicates that they belong to one and the same family. Some children did not like such a conservative family, splitting out from parent's house. These are shown in Fig. 13.17d, e and f.

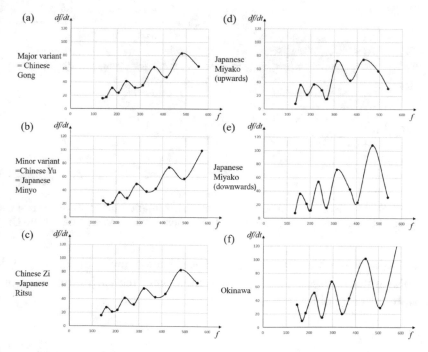

Fig. 13.17 Phase portraits of pentatonic scales: **a** major variant, **b** minor variant, **c** Chinese Zi, **d** Japanese miyako (upward), **e** Japanese miyako (downward), **f** Okinawa

13.4.3 Other Scales

Figure 13.18 depicts phase portraits of three kinds of more artificial scales: a whole-tone scale (a), a chromatic scale (b) and Messiaen scales (c–f). A characteristics of the portraits is their regularity. The portraits of the whole-tone scale and chromatic scale are just linear, monotonically increasing with different gradients. Messiaen scales are regular oscillations. Music with such regular scales could be very boring. But, of course, there are composers who can make the impossible possible, like Nikolai Rimsky-Korsakov, Modest Mussorgsky and Claude Debussy (Sect. 3.2.3).

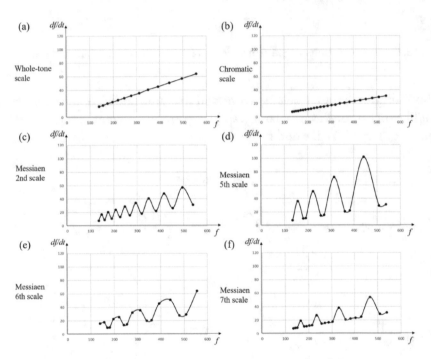

Fig. 13.18 Phase portraits of **a** whole-tone scale, **b** chromatic scale, **c** Messiaen 2nd scale **d** Messiaen 5th scale, **e** Messiaen 6th scale **f** Messiaen 7th scale with a base tone C3

Interestingly, the Messiaen 6th and 7th scales are comparable to some scales with blue notes in Fig. 13.19. However, there is no regularity in blues scales, probably therefore blues is more interesting.

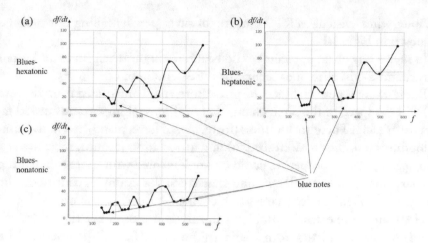

Fig. 13.19 Phase portraits of blues: hexatonic, heptatonic, and nonatonic. Arrows indicate the regions of blue notes

13.5   Concluding Remarks

In this chapter we have applied the phase portrait method to musical pieces and we show what we can see with this method. It might explain our imagination coming from our mind when we listen to or play music in form of graphics.

What is the difference in our feeling between Bach's prelude C major and Gounod's Meditation, for example? Can we see differences between these two portraits? Yes. The curves of these two phase portraits evoke an essential distinction: the former has a rigid backbone, while the latter is floating and flying away as a song without backbone. Such a distinction may correspond to differences in our feeling. Bach's blue curve gives us the impression that we are steadily going down along the backbone, listening to the Meditation we feel as if we are flying up to the sky.

From the phase portraits of the Emperor String Quartet of Haydn we can learn visually about Haydn's intention how four members together as a group should play a theme and the following four variations: who is the central player and how do the other three support the central player. If some of the members do not understand their roles and expose themselves too much, the performance is a disaster, as we can see in Fig. 13.12c.

For the phase portrait of the Guillaume Tell Overture of Rossini we can watch "literally" the scene: after a fanfare a small number of horses start galloping straight ahead, the number of horses increases with spreading in the running direction, and shaking heads…. Also the portrait of Chopin's Etude

"Winter wind" visualizes KT's imaging of silver powder falling down on her, as shown in Fig. 13.1.

In such a way phase portraits would be interesting and helpful to understand music, and to visualize one's feeling or the intention of composers.

But nothing is perfect. Since the phase portrait is constructed from the pitch (f) and the speed of the pitch change (df/dt), several important musical factors are missing. These are loudness (forte, mezzo forte, piano,......), dynamics of loudness (crescendo, decrescendo) and articulation (tenuto, marcato, staccato, legato). These properties could be implemented with more sophisticated software, maybe using different line characters. For example, continuous line for legato, broken line for marcato, thick line for forte, and so on.

There are more essential problems in this approach. Since the values of f and df/dt are calculated from 2 adjacent notes, we have difficulties to define appropriate neighboring notes for the following cases:

1. rest and end
   In both cases the adjacent note is missing.
2. 2 notes or more are played at the same time
   Which note should be taken for the plot?

There are no general solutions for these problems. We have to find reasonable ways case by case. For example, most of the short rests can be treated, as if the former note lasts without rest till the next note is played. Actually, this problem is almost the same as the problem of articulation. For the end note we just finish with the note pair formed from the previous note and the end note.

Concerning the second problem, one may pick up the dominant note among notes played together, or separate them into multiple lines, as if they are played by different instruments. This would differ from piece to piece.

In spite of such difficulties phase portraits can help us to imagine some kind of intrinsic shape which musical pieces possess.

Moreover, very simple phase plots for various scales tell us at a glance which scales belong to the the same family or are closely related, but from different groups. We can notice immediately that all church modes have a similar portrait, just following a shift where to start. However, they are completely different from those of Gypsy scales or maqāms. On the other hand, some of the Indian scales are exactly identical to corresponding church modes or Gypsy scales, suggesting that Indian music has influenced church modes and Gypsy music.

In the same way the pentatonic scales can be easily categorized into some groups according to the portraits. We can also understand why the whole-tone music, chromatic music and Messiaen music are not so exciting. The characterictics of them are very regular.

Before closing this chapter, we like to share our "Aha" experience. KT exclaimed "What???", when she was practicing the piano part of Prokofiev's violin sonata No. 2. There is a passage shown below. What is this? It is certainly not a major scale, not a minor scale, not a chromatic scale...Funny sequence, which is normally never heard....

Then, SCM said "Why don't you make a phase plot?" KT started to calculate, before she answered "Yes, Sir!"

Figure 13.20b is the phase portrait of the mysterious passage, and KT said "Aha!". Look at Fig. 13.20a. Prokofiev's passage is not so far from one of the Indian ragas kaphi.

We do not think that Prokofiev used an Indian scale. But we know now that the passage is somewhat similar to the Indian music.

One of the essential themes which we could not show in our phase portraits is the major/minor problem. Till now we have not found any significant difference between major and minor in this approach. This will be our topic for future projects.

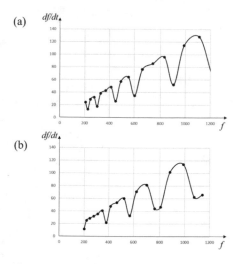

Fig. 13.20 Comparison of **a** one of the raga scales (kaphi) and **b** Prokofiev

References

1. K. Tsuji, Acoustic spirals: Analysis of Bach's prelude in C major, in *Spirals and Vortices - In Culture, Nature, and Science*, ed. by K. Tsuji, S.C. Müller, Nature and Science (Springer International Publishing, Cham, 2019), pp. 113–125
2. H.-G. Schuster, *Deterministic Chaos: An Introduction* (Physik Verlag, Weinheim, 1984)
3. D. Randel (ed), *The Harvard Biographical Dictionary of Music* (Harvard University Press, Harvard, 1996), pp. 1010–1011
4. B. Wiermann, Schwencke, in *Die Musik in Geschichte und Gegenwart*, Vol. 15, ed. by L. Finscher,.(Bärenreiter/Metzler, Kassel, 2006), pp. 438–439
5. E. Heimeran, B. Aulich, *Das stillvergnügte Streichquartett* (Fränkischer Tag GmbH & Co., Bamberg, 1956)

14

What Is Music Supposed to Do with Our Mind and For Our Society?

We visualize in Chap. 13 by using phase portraits many of our imaginations which we obtain by listening to music. But still there are many things which cannot be explained by phase portraits: comfort, satisfaction, happiness, or sadness, …. In this chapter we try to find out where such emotions come from, what we do or expect when listening to certain music, and finally what is the relationship between music and our society.

14.1 Major and Minor

14.1.1 Introduction

Certain aspects of music influence our perceived emotion. Here we focus mainly on mode as an example. So we start the question: What is the difference between major and minor?

Both major and minor scales belong to the church modes, which consist of five whole tones and two half tones in the octave: counting from the tonic tone, the major scale has half tones between the third and the fourth tones and the seventh and the eighth tones, while the minor scale between the second and the third tones and the fifth and sixth tones. Below the C major and C minor scales are compared.

© Springer Nature Switzerland AG 2021
K. Tsuji et al., *Physics and Music*,
https://doi.org/10.1007/978-3-030-68676-5_14

The following brackets are used for indicating the type of interval between subsequent tones:

$\wedge$: minor second (1 half-tone interval)

no sign: major second (1 whole-tone interval)

⌐⊥⌐ : minor third (1 ½ whole-tone interval)

⌐‾⌐ : major third (2 whole-tone intervals)

It is often stated: pieces in the major key sound cheerful, happy or conclusive, while those using the minor key sound sad, unhappy or inconclusive. Is this true? And if so, why?

We notice the difference between major and minor more clearly, when the melody undergoes a change from major to minor or from minor to major. A typical example is heard in "The Lindenbaum" of Franz Schubert (see the notes in Fig. 3.3 in Chap. 3).

The text of Wilhelm Müller starts with

Am Brunnen vor dem Tore,	There stands a Lindenbaum
Da steht ein Lindenbaum;	at the fountain in front of the gate;
Ich träumt in seinem Schatten	In its shadow I used to dream
So manchen süßen Traum.	some sweet things.

which is a just happy statement. For this part Schubert composed a melody in E major.

On the other hand, the text in the middle

Ich mußt auch heute wandern	I had to travel passing the tree
Vorbei in tiefer Nacht,	during the deep night.
Da hab ich noch im Dunkel	Although it was dark,
Die Augen zugemacht.	I closed my eyes

does not describe any happy situation but somehow a more problematic one. This part is in G minor.

So far we can understand the different feelings of major and minor very well: a happy situation with major melody and a problematic situation in minor.

But this does not apply to all cases. Read below the Japanese poem translated into English:

A Happy Doll's Festival (by Hachiro Sato)
    Let's light the paper lantern
    Let's give peach flowers to dolls
    Five doll-musicians play flutes and drums
    Today is the pleasant Doll's Festival

The melody for this poem composed by Koyo Kawamura:

A happy doll's festival

Isn't this melody in minor? The scale of this piece is pentatonic. KT sang this song often in the Kindergarten. She never felt any sadness in it. But, the Japanese doll's festival is not cheerful, but rather mellow.

As we have know from Chap. 3, there are the major and minor variants for the pentatonic scale. The major variant has 1.5 intervals between the third and fourth tones and the fifth and sixth tones, while the minor variant has these between the first and the second tones and the fourth and fifth tones:

C major variant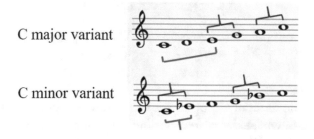

C minor variant

It is not so clear whether the major variant sounds happier than the minor variant. Try to compare the three pentatonic melodies shown in Chap. 3: "Laterne" (major variant), "Morning Mood" (minor variant) and the 2nd movement of the Symphony No. 5 of Antonín Dvořák (minor variant).

There are also the major and minor Gypsy scales:

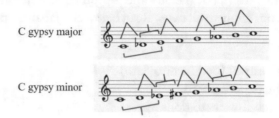

For both authors (a Japanese and a German) it is difficult to say whether Gypsy major is happier than Gypsy minor or vice versa. The important is: As indicated by the blue brackets, all three major scales have a major third, while all three minor scales have a minor third at the beginning.

We are sure that mode has an effect on our perceived emotions, but how exactly it does or does not have a clear-cut answer. In the following, we go through some aspects in more detail.

14.1.2 The Major Triad and the Minor Triad

Before we discuss more about the relationship between the mode and our emotion, we look at the triad (a harmonic set of three notes, see Sect. 3.1.2.4), which directly connects to the mode. The root (C) and the fifth (G) are common for C major and C minor scales. Only the thirds are different: the major triad with major third (E) for the major scale, and the minor triad with minor third (E♭) for the minor scale:

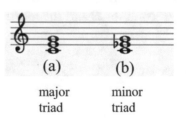

The harmonic series of C2 up to the 6th harmonic comprises C2, C3, G3, C4, E4 and G4, where the corresponding frequency ratios are 1:1, 2:1, 3:2, 4:3, 5:4 and 6:5 (see Table 4.3 in Chap. 3). All three tones of the major triad, C4, E4, and G4 are included in the harmonic series of C2. Therefore, the major triad sounds consonant. On the other hand, the minor triad contains E♭4, which is a half tone lower than the 5th harmonic, E4. As a result the minor triad sounds somewhat mismatched.

Many music theorists and composers worked on the minor triad or minor scale involving the harmonic series. Here are some references [1–4].

Further, concerning the major and minor triads, three major triads (or inversed ones) are derived from the augmented triad, when one of the three notes is lowered by a half tone. In a similar manner three minor triads are derived, when one of the three notes is raised by a half tone. These are illustrated in Fig. 14.1 for the case of the augmented triad with the root C.

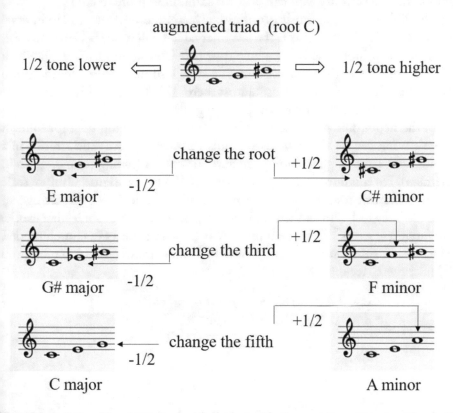

Fig. 14.1 Relationship of the augmented triad with the root C and triads with a half tone falling or rising indicated by -1/2 and +1/2, correspondingly

Norman Cook advocates a close relationship between the falling pitch to major/the raising pitch to minor and animal vocalization [5]. This is from the observation that animals who try to show their strength tend to use a low or falling pitch [6]. His opinion is fresh and worth noting. Yet there remains a large space for further discussion.

14.1.3 Finished or Unfinished

Besides eliciting different emotional valence, properties of the major and minor scale also lead to other feelings, for instance whether we perceive a musical piece as finished, or to what degree we feel comfort.

If you sing or play an octave of the heptatonic C major scale, you finish singing/playing with satisfaction. If you sing an octave of the heptatonic C minor, on the contrary, you will wait for something more to come.

The difference of this finished/unfinished feeling comes from the last interval in the scales. In the major scale the last interval is a half tone (B-C in C major), while in the minor scale it is a whole tone (Bb-C in C minor). The B, which is a half tone lower than the last tone, is a tone leading towards the end. John Powell describes this situation as "almost there" [7]. Yes, with this leading note we know that we have almost come to the goal.

In the heptatonic minor scales there is no such a leading note for "almost there". What can we do, if we like to finish a musical piece satisfactorily, in a minor key? To get a good ending we have to create a leading tone (Leitton in German). For this purpose one uses B instead of Bb in the original C minor scale (natural minor), resluting in the harmonic minor scale. With this change, at least one may feel "almost there". However, this scale sounds somewhat exotic, because there exists a 1.5 tone interval which never appears in church modes. The melodic minor scale is a compromised product between church mode and "almost there" feeling.

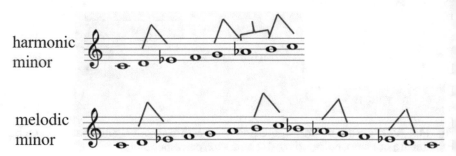

For the Gypsy scales (minor and major) the last interval is also a half tone. However, it does not give us the feeling of "almost there". Although the last four notes of Gypsy scales are identical to the harmonic minor scale (G-Ab-B-C), we cannot reach the goal with the Gypsy scales. Interestingly, however, Gypsy scales do not supply us with an unfinished, unsatisfied feeling like a natural minor scale, but provide the impression of never-ending. With Gypsy scales we can continue dancing and jumping. Maybe the reason for this is the repetition of three successive intervals 0.5–1.5–0.5.

Both major and minor variants of pentatonic scales do not have any half tone. However, some pentatonic melodies provide us a feeling of finished and others of unfinished. A happy doll's festival belongs to the former case and the German children song "Laterne" (see Chap. 3) to the latter case. They really depend on the type of composition. It is difficult to find a general rule.

It should be noted that many modern music pieces end without leading tone, making the piece more interesting—with an unfinished feeling or expectation for something more to come.

14.1.4   Comfort or Discomfort

The tritone (an interval encompassing three whole tones) is usually called "devil's interval". The reason is the fact that when listening to this interval we feel uncomfortable, uneasy or dark.

In Sect. 14.1.2 we explained why the minor triad sounds somewhat dissonant—E♭4 is mismatched with E4 (the fifth harmonic of C2). The mismatch in the tritone is much worse. The tritone with the root C4 is C4-G♯4 where the G♯4 has conflicts with both the third harmonic G3 and the sixth harmonic G4. This is not the level of "somewhat mismatched", but "very strongly dissonant".

Blood et al. [8] studied neurophysiological correlations between the sequence of harmonics and perceived emotion (comfort/discomfort). When the level of the dissonance becomes higher, a feeling of discomfort for the test persons increases. Such discomfort correlates with the activation of paralimbic and neocortical brain regions that have been associated with emotion processing.

In particular, comfortable emotional experience at hearing music is called "chills" (in German "Erschauern" or "Gänsehaut"). During "chills" the cerebral blood flow increses in the region where reward anticipation and emotion are processed [9].

Moreover, the relationship of consonance/dissonance to comfort/discomfort feeling seems to be independent of the cultural background [10]. The Mafa people in north Kamerun who have never heard European music also experience dissonance discomfort.

Coffee Break

Nevertheless, dissonances are used in many musical pieces. Adding a small amount of salt makes cakes sweeter.

14.1.5 Happy, Sad or High

We now come back to the question at the beginning: Is it true that pieces in the major keys sound cheerful, happy or conclusive, while those in minor sound sad, unhappy or inconclusive? Who claims this? Does this depend on one's cultural background? Have we learned so?

John Powell writes that a three-year old boy mentioned the first part of Rachmaninoff's Second Symphony (in a minor key) as being "sad" [7]. We believe that. Powell says also that the statement "major key: happy/minor key: sad" is not an absolute rule. Not only melody and harmony but also other physical factors like rhythm, tempo and loudness may strongly affect the reception of a musical piece. Rachmaninoff's Second Symphony is not only composed in a minor key but also lethargic (largo) and heavy with deep tones of the double bass or trombone.

Dalla Bella et al. [11] studied effects of tempo and mode with adults and children aged 3–8 years. Their results show that adults and elder children (6–8 years old) can notice the difference of mode, as well as tempo, 5-year-old children can distinguish only the change of tempo, and children of 3–4 years age can distinguish neither the change of mode nor of tempo.

Let us discuss influences of various physical factors with some examples.

14.1.6 Boléro

Boléro is a Spanish dance music with rather slow tempo. The famous "Boléro" of Maurice Ravel was originally composed as a ballet assigned to Russian dancer Ida Rubinstein. This is a unique piece with a characteristic rhythm (see below), maintained consistently from the beginning to the end by a snare drum (Fig. 14.2). We are sure that most of the listeners become "high" towards the end. Maybe therefore this piece is played often in New Year's concerts.

The Boléro consists of the two groups of melody A and B, as shown below, and only of these two melodies except those for the coda. It goes AA BB AA BB AA BB AA BB A B and coda.

Fig. 14.2 Drum beats repeated 169 times without interruption (photo by KT)

Melody A and melody B of the the Boléro

Are these melodies major or minor? Part A can be played with only the white keys of the piano. It starts with C5 and finishes with C4. But this does not necessarily signify C major. It could be A minor or another church mode. What we can say is that it is heptatonic with 5 whole-tone intervals and 2 half-tone intervals, that is, a church mode. For ears familiarized with European music, the melody A sounds neither major nor minor. It sounds a little bit exotic—provocative dullness, maybe....

Part B consists of the series D natural minor, D harmonic minor, F natural minor, D natural minor, and F natural minor. On the basis of natural minor the harmonic minor has a special effect. Raising a half tone here is like a sun beam penetrating the clouds. This effect disappears again in the cloud of natural minors with decreasing pitch.

After repeating A and B several times, at the coda, the key changes from F minor to F♯ minor—a half tone higher. This is surprising for listeners after hearing repetitive melodies in the same keys, attracting their attention and giving a fresh impression. So we can say that a sudden change of the key towards the higher frequency in Part B and in the coda contributes to a somewhat brighter feeling. However, this is not reason enough yet for putting "Boléro" on the list of a New Year's concert.

How about the speed? Ravel indicates 72/min at the beginning, and later no tempo indication appears any more. Thus the tempo is constant throughout the piece. The drummer should hit 7.2 times per second during the triplet.[1] Repeating the same melodies with constant tempo. Is it not boring?

When the authors compared some different recorded performances, however, the performers have a tendency to increase the tempo towards the end. Some start with 72/min and finish with 76/min. Others start with 76/min and finish with 80/min. The acceleration of tempo supplies an active feeling.

An interesting (and may be essential) fact is that the piece is played by human beings and not by a machine. It is played with "almost" constant speed but not exactly (in another word, a little bit fuzzy). There is some deviation in tempo depending on the solo instrument who is just playing the melody, and there are also slight phase shifts. (Players have to breath occasionally!) These variations in tempo and phase shifts are readjusted in the period between one melody and the next melody, so that all instruments can, in principle, play in the same tempo and the same phase throughout the piece. And to realize this, a real mastership of the conductor is required. Such interplay between fuzziness and order attracts the attention of listeners. As we mentioned in Chap. 5, African fuzzy rhythms make their dance more attractive. Remember that Boléro was originally composed for dance.

---

[1]Tempo 72/min = 1.2/s. One quarter note consists of 2 times 16th triplets (in total 6 notes). Therefore, the speed of the triplet playing is $6 \times 1.2/s = 7.2/s$.

Furthermore, the loudness increases almost monotonically with increasing number of instruments. It starts with the tambour beating the characteristic rhythm with **pp**, accompanied by pizzicato of the viola and cello. And then the first flute starts playing the melody. Soon, the second flute joins in rhythm, the clarinet replaces the flute for the melody with **p**. In the middle of the piece the rhythm is beaten with **mf** by a number of additional instruments. Shortly before the end most of the orchestra instruments are involved, all of them playing with **ff**. Finally, for the coda, the basic key is raised by a half tone. Tempo is raised. All members including cymbals and gong play together with **ff**.

This is a happy start for a New Year.

14.1.7   Siegfried Idyll

Richard Wagner composed the Siegfried Idyll as a present to his wife Cosima, when his son Siegfried was born in 1869. The original composition was for a small chamber orchestra of 13 members: flute, oboe, two clarinets, trumpet, two horns, bassoon, two violins, viola, cello and double bass.[2]

In contrast to Boléro we prefer to listen to the Siegfried Idyll alone at home. The following melody in E major comes quietly, and the tempo is very slow:

The first theme of the Siegfried Idyll

This introduction is the expression of Wagner's happiness from within. The key of this melody shifts from E major to B major, C♯ minor, transition with some dissonances, and comes back to E major. Emotion increases with the higher pitch (from E major to B major). Wagner was very, very happy. He wished to occupy this happiness alone. He was so happy that he was afraid to lose it (C♯ minor and the following dissonances). Yet he was happy (coming back to E major)....

---

[2]Later, for financial reasons, Wagner rewrote this piece for a larger orchestra and sold the score.

The motif above appears in many places in this piece, expressing his happiness which is almost bubbling over, and embarrassment to show this. In between, there are changes in keys, changes in modes, dissonant accords, and repeating triplets without changing the pitch, showing fluctuations between his happiness and anxiety. It is a daydream.

Suddenly he snaps out of his daydream. This is not a dream. My son, Siegfried is there!

The second theme of the Siegfried Idyll

Then, the first motif and the second motif resound alternately or in parallel. It is a mixture of dream and reality, happiness and anxiety. And finally everything is faded out......in a sweet dream...*Rallentando, pianissimo*...the happiest dream....

14.1.8   Swan Lake

Swan Lake, Op. 20, is a ballet composed by Pyotr Ilyich Tchaikovsky in 1875–1876. The story may be from a Russian or German fairy-tale and there are many variations. The following is an example:

Princess Odette and her maidens are condemned to be swans. They change back to humans for a few hours during the night. Only a true love can break the spell. The prince Siegfried, who is forced to select his wife, sees Odette in the night at the lake and falls in love. At a ball, however, Odile, the daughter of the sorcerer, appears and the prince misunderstands that she is Odette. He vows to love Odile, so that there is no chance for Odette to move back to be human. In despair, the prince and Odette throw themselves into the lake (Fig. 14.3).

Fig. 14.3 Ballet scene in the Swan Lake, a 2008 production at the Royal Swedish Opera (CC BY 3.0 by Alexander Kenney/Kungliga Operan)

The famous melody is played in the second act, as the Scene Finale:

The theme melody of the Swan Lake

This is a very emotional and wailing melody in B minor. The accords for the orchestra score are also all in minor. Everybody agrees that this part indicates the tragedy of the couple. The crescendo to fortissimo towards the end brings along the climax of the tragedy.

Here we find a typical minor scale with a feeling of sadness.

However, if the tempo is changed from 84/min to 160/min, it does not sound so sad. Or if a drum beats the following rhythm as a background,

it does not sound tragic either. It is almost like rock music. Sorry for Tchaikovsky, but we like to emphasize that not only the scale and harmony of minor/major but also the speed and rhythm are very important for our feeling.

Summarizing the Theme "major and minor"

Often music with a major key provides happy feelings, while that with a minor key expresses problematic situations. A typical example is "The Lindenbaum". But some musical pieces with a minor key are not necessarily sad or unsatisfactory. Rhythm, tempo, tempo change, key change, …all these factors contribute to our perceived emotion, as is seen by the examples of Boléro, Siegfried Idyll and Swan Lake.

> **Coffee Break**
>
> KT once experienced a nightmare with one (not so well known, yet professional) philharmonic orchestra. They managed to mess up one of the most beautiful pieces of Mozart—the second movement of the Piano concert No. 21 in C major (see below) with inappropriate beats of triplets: strong weak weak, strong weak weak. A devil is dancing …
>
>

14.2 Music and Our Society

Music evokes strong emotions. We now look at the role of music, including its role in transmitting emotions, in a broader societal context.

Music, when performed with its elements of rhythm, melody and harmony has always motivated people to sing and move, be it in ceremonial processions in large events or individually by dancing.

Together with our human history music was born in many places from religious environments. People sang and danced for the purpose of worshiping their gods. At weddings or the birth of children they sang happy songs, and at funerals they sang sad songs. In this way music offers a fundamental way for sharing emotions in the society.

14.2.1   Music and Moving

We learned from Chap. 11 what our hearing organs do and how the signals are processed for understanding what we perceive upon hearing music (though there still are a lot of black boxes). Furthermore, perceiving music also affects the motor regions of our brain, and by consequence affects our movements. Now we will show some examples of our body reactions on hearing music that results from such signal flows.

When a pianist listens to a piano piece, not only the auditory nerve system but also neurons in the motor system are activated, although he/she actually does not play piano. In another case, when a pianist sits in front of a piano and only imagines to play a piece (that does not create any sounds), his auditory system is activated as if he would really play it [12]. Pianists know how they should move their fingers to play the piece, and they also know how the music should sound when they perform certain movements of their fingers. Similar phenomena are observed for musicians who play other instruments. However, this does not happen for people who have not played any musical instruments. The brain's ability to activate specific motor representations upon perceiving or imagining instrumental music is thus established by training.

However, there are some aspects of music that are not tied to a specific instrument or mode of performance. For instance, the premotor cortex of non-musicians was activated in response to auditory rhythmical stimulation [13].

Therefore, if one hears music in which the rhythm is dominant, the body reacts immediately. In addition, by listening to melody and harmony, the body can express happiness, sadness, passions, loneliness due to what one feels while listening (see Sect. 14.1). This is a dance. Dancing upon hearing music is, in this way, a very natural reaction.

But then, dance without music is difficult to imagine. In the famous film "Amadeus"[3] there is a dancing scene without music in a newly composed opera. The reason for such a performance was that Emperor Josef II forbade Mozart to use music during dancing according to a nasty suggestion by Antonio Salieri. Mozart obeyed the Emperor and took music away from the dancing scene. When the Emperor saw the dance performed without music, he felt very strange and immediately allowed Mozart to use music again. Not only Emperor Josef II but most of us agree that music is an essential element of dance.

Different from operas, the importance of dance in musicals is equivalent to that of singing. Musical singers should be able to dance and dancers should be able to sing. Figure 14.4 shows a dancing scene in a performance of West Side Story.

Fig. 14.4 Dance scene in West Side Story, Shark girls in "America", Portland Center Stage production, 2007 (public domain, photo by Owen Carey)

In some sports music is also involved. Among various disciplines of artistic gymnastics the female floor exercise is accompanied by instrumental music, probably because there is an essence of ballet in this discipline. The same is true for the rhythmic sportive gymnastics. For a competition in figure skating, the selection of the music is very important. Synchronized swimming is accompanied by music, as well, although swimmers cannot hear the music so much when their heads are under water. In Fig. 14.5 some nice scenes are presented. Maybe some of you can hear the music when watching these pictures.

Beyond performances such as ballet, musical or artistic gymnastics music is also often used in order to let people move together, to the same direction

---

[3] Amadeus is a 1984 American film directed by Miloš Forman and adapted by Peter Shaffer from his stage play Amadeus.

Fig. 14.5 Left: Aliona Savchenko/Robin Szolkowy at the 2011 European Figure Skating Championships (CC BY-SA 3.0 by David W. Carmichael), middle: Wu Yiwen and Huang Xuechen of China perform during the duet technical routine at the 2013 French Open synchronized swimming (CC BY-SA 3.0 by Pierre-Yves Beaudouin), right: Nadia Comaneci performing a floor exercise at the 1976 Summer Olympics (public domain, Comitetul Olimpic si Sportiv Roman)

or with the same speed. This is exemplified by musics for folk dances, social dances and military marches.

Intermezzo

A music that never works as intended (to make people sing together) is the accompaniment by an organ in churches.

   KT is not a Christian. But since she lives in Germany, she occasionally goes to church for a wedding, a funeral, or other celebrations of her friends or relatives. She has no problem to visit a church, so far as a church organ is not played too loud, which does not mean that she dislikes organ music. When organs are played "as it should be", she likes them. It is still acceptable when the loud organ ends like phasing out quietly. Unfortunately, however, it often happens that the sound of the organ is consistently very loud till the end. Especially, when people sing, the "accompanying" organ gets louder and louder, as if the organ wants to teach people how to sing. Since people hear the organ and then sing, the sound of the song is delayed. As a consequence, the organist plays even louder. This does not sound like good music. In this way the sound of an organ can become very insisting. This is the best way to stop liking church music.

14.2.2   Musical Expectation and Moving

Music has an enormous suggestive power, it may induce interactive reactions leading to sensible behaviors or spontaneous movements.

Let us get to an aerobics studio. People come to the studio for fitness exercise. Usually music here is very loud. Sometimes it is too loud to hear what the instructor says. But it does not matter. Owing to the music most of the participants move their body spontaneously, rhythmically and synchronously following the instructor. Actually they do not listen to the instructor, but to the music during exercising. They know (guess) what comes next in the music, and the body knows the next movement at the same time. However, there are a few people moving their body with some delay. These people tend to learn the sequence of movements from what the instructor said without listening to the music. They do not get important information from the music (what is the next movement?), and therefore, there is probably a time delay. If you have to recall the sequence of movement beforehand: raising the right hand while turning left,…or maybe to the right, or raising the left hand instead…stumbling over your own feet……

Our next station is a riverside to watch a regatta. An eight-men regatta consists of a boat with nine persons: eight rowers and a coxswain, as shown in Fig. 14.6. In order to synchronize the movement of eight rowers the coxswain gives signals. No metronome could substitute the coxswain, because he does not convey a constant tempo. He has to give the right tempo at the right time. He watches his competitors, how far they are behind or in front of him, he has to watch his eight rowers to estimate how much energy they still have,

Fig. 14.6 Men's eight of Harvard University in Rowing at Henley Royal Regatta 2004. Taken place on 1st of July, 2004. The coxswain sits at the bow of a ship facing 8 rowers (public domain by Guety)

and then he has to decide how fast and how long they can row. And the eight rowers should react to his signal immediately and synchronously.

If rowers listen to the coxswain and then react, the body movement for rowing could have a certain delay, which is similar to the case of some people in the aerobics studio. Also the reaction time from listening to moving is slightly different for each person. In the worst case each of the eight rowers fights against the other seven rowers instead of cooperating. The stronger they row, the lower the speed becomes.

A coxswain of an eight-men regatta does not sing a song. But the way how to give signals is, we can imagine, rhythmical like a song, the "song" which the eight rowers hear at every training. Therefore, rowers can foresee what comes next, and they can realize their movement together with the coming command. In this way a good teamwork of all nine members is established.

14.2.3   Two-Edged Blade

With music we can enjoy watching operas, ballets, musicals, as well as sports. We can also enjoy dancing together with a partner or with many people. Our body training can be more efficient with music. With music everything looks fine so far.

However, we should not forget that music is a two-edged blade, similar to the natural sciences. Authors, like scientists, have to watch whole the time for which purposes scientific and technical achievements are used. Otherwise, needless to mention ABC weapons and other global menaces, many outcomes could induce an environmental disaster or human diseases. If we would not make similar considerations in the field of music, music could be also dangerous.

March music is appropriate, for instance, for an opening ceremony of the Olympic games. All participants can walk with a pace given by march music, presenting a well-organized discipline. Military parades of walking soldiers are much better synchronized with military march music.

Just imagine that all people are walking together—left, right, left right—without thinking where to go but straight ahead. Be careful. Some music has a charismatic power to let people behave collectively without thinking whether it is right (left???) or not. Such kinds of music buoy up soldiers' spirits. They could march to death without being afraid.

Mass games were and are shown often in order to emphasize the power of states, dictators, or organizations. Here all participants of large number perform choreographic movements (dance, march, gymnastics, …), not as an individual but as a tiny element of a total group, often with appropriate music.

Coffee Break

Regardless of with or without music, "to stop thinking" is never good. To follow an automatic car navigation system without thinking can lead you to a very dangerous situation. "Turn around!" on a highway may be fatal: Such a suggestion often comes from a not up-dated software. "Drive straight!", and then suddenly you fall into the water. The car ferry which you plan to take has not yet arrived at the harbor in front of you……

A sensitive case is a situation when musical compositions radiate to political conditions and opinions. Honestly speaking, the authors had and have big psychological problems to write about the music of Richard Wagner. He was the person who wrote the "Judaism in Music" (1850). In this antisemitic essay he attacked Jews in general, Jewish composers like Giacomo Meyerbeer and Felix Mendelssohn Bartholdy. He criticized their music as being "shallow" and "artificial". On the contrary, his works are, he himself stated, genuine and possess the "German spirit". About 80 years later his music, especially "great" operas became a symbol of the national socialism of Hitler.

We know some people, among our friends, who have not and will not listen to "Wagner". In Israel Wagner's works were not played for a long time. Famous Jewish conductors like Zubin Mehta have tried and failed to play Wagner in Israel. Good music should be played! Daniel Barenboim and Kan Kol Hamusica also have the same opinion, tried and failed. Even after broadcasting a part of Wagner's "Götterdämmerung" (Twilight of the Gods) on radio in 2018 there were strong protests in Israel, and the radio station apologized.

We find some of Wagner's compositions very impressive. Can we simply enjoy such musics? Or should we not? No, this is not a matter of "being allowed" or "being not allowed". To criticize Wagner and to criticize Wagner's music are two different things. The "Siegfried Idyll" is too beautiful to be ignored. It touches our heart deeply. The "Tannhäuser", the "Flying Dutchman", and other operas are also worth listening to. At the same time Mendelssohn's music is not shallow or artificial as Wagner wrote. It is you who listens to music, and it is you who is impressed or not impressed by various facettes of music. Be honest to yourself and trust your own feeling.

14.2.4 Music-Related Illness and Therapy

Since illness is too far from the authors' expertise, we just show one example for music-related illness, focal dystonia and a brief explanations about application of music therapy.

14.2.4.1 Focal Dystonia

Some pianists suffer from focal dystonia. When they intend to play piano, some of their fingers do not react as they should. For example, when the index finger should push a key, the middle finger also moves simultaneously, resulting in unwanted noise creation. The more they exercise, the worse the performance becomes.

This is a neurological condition, where muscles in certain places are contracted involuntarily, or positioned abnormally. One of the most probable explanations is the misfiring of neurons in the sensorimotor cortex (a part of the motor cortex). It is caused due to overlapping of the areas which are in charge of each finger movement [14, 15]. Since musical training shapes structural brain development [16], it could happen sometimes (not always) for somebody (not everybody) that the structural brain development proceeds too far by hard training.

Similar phenomena occur also for musicians who play instruments other than piano, as well as in sports.

14.2.4.2 Music Therapy

Music can be used for therapeutic purposes. Music therapy is an evidence-based therapeutic method. Since music has various effects on our minds, body and society, some of these effects can be applied to patients to improve their mental as well as physical conditions.

Music has positive effects on motor and attention skills, as well as socio-communication and interaction skills, characteristics which can, for instance, be used in therapies for patients with autism spectrum disorder. For stroke patients music can help recovering motor skills. Patients suffering from dementia who join to sing or play instruments may obtain feelings of enjoyment, awareness, and engagement.

For details about music therapy, there are some good books, for example, "The Handbook of Music Therapy" [17] or "A Comprehensive Guide to Music Therapy: Theory, Clinical Practice, Research, and Training" [18].

Towards the end, we like to present one very nice classical example: singing for speech problems. The movie "The King's Speech" (2010)[4] shows how the future King George VI learned and exercised to speak without stuttering. A famous scene is that the Australian speech therapist Lionel Logue "conducted" in front of the King, when the King had to speak for the radio broadcast on Britain's declaration of war on Germany in 1939.

The preceding chapters and some others provide evidence about the intense and often intimate relationships between the major issues of this book: physics and music. They also demonstrate the relevance of many other disciplines woven into and connecting the main topics concerned: physiology, neural dynamics, psychology, philosophy and social science.

May the collected material serve as a motivation for the reader to dive into greater depths for exploring this rich realm of a fascinating theme.

14.3 Bye Bye, I Must Go

Recently, the authors should have stayed at home for the whole time because of the measures against the corona virus (covid-19). Currently restaurants and bars are closed. There is only a limited number of sport events, theaters and concert halls are closed.

It is amazing that homes are still full of music. Some concert houses or theaters offer their concert videos to be played at home without charge. Some musicians organize live concerts on the internet: Each member of an orchestra plays alone at home according to the instructions of a conductor and the individual sounds are collected. There is a nice song of May Raabe "Today I did nothing. ......I am staying at home, and what I do: I open the door of the refrigerator and close it again,..." (original in German).

Music helps us to stay at home without physical contacts. Musicians come and go, but music stays in our hearts.

Bye bye, I must go.

When shall we three meet again?
In playing drum, trumpet or violin?

(Three musical witches as metamorphose from Shakespeare's Macbeth)

---

[4]The King's Speech is a historical drama film directed by Tom Hooper and written by David Seidler.

Bye bye, I must go.

Kinko made a sketch from
a photograph „Gypsy boy with cello"
by Eva Besnyö

References

1. M. Hauptmann, *Die Natur der Harmonik und Metrik* (Breitkopf und Härtel, Leipzig, 1853)
2. A. von Oettingen, *Harmoniesystem in dualer Entwicklung - Studien zur Theorie der Musik* (Dorpat und Leipzig, Leipzig, 1866)
3. S. Karg-Elert, *Polaristische Klang- und Tonalitätslehre* (Leuckart, Leipzig, 1930)
4. P. Hindemith, *Unterweisung im Tonsatz. I. Theoretischer Teil* (Schott, Mainz, 1937)
5. N.D. Cook, *Harmony, Perspective and Triadic Cognition* (Cambridge University Press, Cambridge, 2012), pp. 26–119
6. E.W. Morton, On the occurrence and significance of motivation-structural roles in some bird and mammal sounds. Am. Nat. **111**, 855–869 (1977)
7. J. Powell, *How Music Works* (Little Brown and Company, Boston, 2010)
8. A.J. Blood, R.J. Zatorre, P. Bermudez, A.C. Evans, Emotional responses to pleasant and unpleasant music correlate with activity in paralimbic brain regions. Nat. Neurosci. **2**, 382–387 (1999)
9. A.J. Blood, R.J. Zatorre, Intensely pleasurable responses to music correlate with activity in brain regions implicated in reward and emotion. P. Natl. Acad. Sci. USA **98**, 11818–11823 (2001)

10. S. Koelsch, T. Fritz, D.Y. von Cramon, K. Müller, A.D. Friederici, Investigating emotion with music: an fMRI study. Hum. Brain Mapp. **27**, 239–250 (2006)

11. S. Dalla Bella, I. Peretz, L. Rousseau, N. Gosselin, A developmental study of the affective value of tempo and mode in music. Cognition **80**, B1–B10 (2001)

12. M. Bangert, U. Hacusler, E.O. Altenmüller, On practice: how the brain connects piano keys and piano sounds. Ann. NY Acad. Sci. **930**, 425–428 (2001)

13. J.L. Chen, V.B. Penhune, R.J. Zatorre, Moving on time: brain network for auditory-motor synchronization is modulated by rhythm complexity and musical training. J. Cogn. Neurosci. **20**, 226–239 (2008)

14. W. Bara-Jimenez, M.J. Catalan, M. Hallett, C. Gerloff, Abnormal somatosensory homunculus in dystonia of the hand. Ann. Neurol. **44**, 828–831 (1998)

15. O. Granert, M. Peller, H.-C. Jabusch, E. Altenmüller, H.R. Siebner, Sensorimotor skills and focal dystonia are linked to putaminal grey-matter volume in pianists. J. Neurol. Neurosurg. Psychiatry **82**, 1225–1231 (2011)

16. K.I. Hyde, J. Lerch, A. Norton, M. Forgeard, E. Winner, A.C. Evans, G. Schlaug, Musical training shapes structural brain development. J. Neurosci. **29**, 3019–3025 (2009)

17. L. Bund, S. Hoskyns (eds.), *The Handbook of Music Therapy* (Routledge, London, 2002)

18. T. Wigram, I.N. Pedersen, L.O. Bonde, *A Comprehensive Guide to Music Therapy: Theory, Clinical Practice, Research, and Training* (Jessica Kingsley, London, 2002)

Correction to: Music Analysis with Phase Portraits

Correction to:
K. Tsuji et al., *Physics and Music*,
https://doi.org/10.1007/978-3-030-68676-5_13

The original version of the book was inadvertently published with an incorrect equation (13.1) on page 364 in chapter 13. The chapter and book have been updated with the following change:

from

$$y_p = y \cdot \sin \alpha + z \cdot \cos \alpha$$

to

$$y_p = y \cdot \cos \alpha + z \cdot \sin \alpha$$

The updated version of this chapter can be found at
https://doi.org/10.1007/978-3-030-68676-5_13

© Springer Nature Switzerland AG 2022
K. Tsuji et al., *Physics and Music*,
https://doi.org/10.1007/978-3-030-68676-5_15

Index

© Springer Nature Switzerland AG 2021
K. Tsuji et al., *Physics and Music*,
https://doi.org/10.1007/978-3-030-68676-5

Printed in the United States
by Baker & Taylor Publisher Services